An Explorer's Guide to the Earth System

D0061016

Ellen P. Metzger

San José State University

Prentice
Hall

Pearson Education, Inc.
Upper Saddle River, New Jersey 07458

Senior editor: *Patrick Lynch*
Senior marketing manager: *Christine Henry*
Production editor: *Pine Tree Composition*
Executive managing editor: *Kathleen Schiaparelli*
Assistant managing editor: *Beth Sweeten*
Art director: *Adam Velthaus*
Cover design: *Daniel Conte*
Cover photo: *Hoodoo at Bryce Canyon National Park (photo by Carr Clifton)*
Assistant manufacturing manager: *Michael Bell*

© 2003 by Pearson Education, Inc.
Upper Saddle River, New Jersey 07458

Printed in the United States of America

10 9 8 7 6 5 4 3 2 1

ISBN 0-13-093335-X

Pearson Education Ltd., *London*
Pearson Education Australia Pty, Limited, *Sydney*
Pearson Education Singapore, Pte. Ltd.
Pearson Education North Asia Ltd. *Hong Kong*
Pearson Education Canada, Ltd., *Toronto*
Pearson Educación de Mexico, S.A. de C.V.
Pearson Education—Japan, *Tokyo*
Pearson Education Malaysia, Pte. Ltd

In loving memory of my mother, Barbara J. Pletcher.

Acknowledgments

I sincerely appreciate the thoughtful commentary of several anonymous reviewers whose suggestions made this workbook much better. I would also like to acknowledge Patrick Lynch, Senior Editor for Geology at Prentice Hall, whose advice and supportive guidance contributed greatly to this project. The production team, led by Patty Donovan, was also most helpful. Thanks to my students and to the teachers of the Bay Area Earth Science Institute, who make teaching about Earth fun. And last, but definitely not least, I am grateful to my husband David for his love, unflagging support, and patience during preparation of this book.

Contents

Module 4: Atmosphere   4-1

Module 5: Cosmosphere   5-1

Glossary   G-1

References   R-1

Preface

Goals

An Explorer's Guide to the Earth System is a student workbook designed to accompany and complement the text *Earth Science* by Tarbuck and Lutgens. The purpose of this workbook is to facilitate infusion of an Earth system approach into introductory Earth science courses and to engage students in active inquiry about their home planet.

To the Student

Earth is your home, most likely the only one you will ever have. Everyday you depend on its air, water, and mineral resources for your survival. Although you may never have thought about it, you are part of a complex system of interacting parts known as the Earth system. Because you are part of the biosphere (living things), other organisms in the biosphere, the atmosphere (air), the hydrosphere (water), and the geosphere (solid Earth) have an impact on you. The intent of this workbook is to supply readings, questions, and exercises to help you gain a greater appreciation for your home planet and its inhabitants. Hopefully, in the process, you'll acquire skills that you can apply elsewhere and develop a curiosity that will keep you attuned to Earth long after this course is over.

Perhaps you feel that you aren't good at science, or maybe you just plain dislike it. It may be that you wouldn't be taking an Earth science course at all if not for your school's science requirement. Try to keep an open mind. You might be surprised by how relevant Earth science is to your life and community. As you explore your own world through study of dynamic and timely topics such as volcanoes, global warming, solar storms, and the search for life beyond Earth, you may surprise yourself. You may get "hooked."

To the Instructor

Today's instructor is confronted with two critical changes in research and education that significantly influence the way that Earth science is taught: 1) The growing importance of a system paradigm that emphasizes interfaces between and interactions among land, water, air, and living things; and 2) Recommendations for improving science teaching through incorporation of active, inquiry-based instruction that lets students take responsibility for their learning and highlights the relevance of science to their lives. Because introductory Earth science courses typically cover a lot of information in a short period of time and most students who take them are non-science majors, it is a challenge to present both fundamental science content and employ a systems approach that assumes knowledge of Earth's subsystems and involves analysis of complex interactions among them. Furthermore, such classes are typically taught to large numbers of students in a lecture format that challenges the ability to engage students in active investigation as opposed to passive assimilation of course material. *An Explorer's Guide to the Earth System* is designed to help meet these challenges

An Explorer's Guide is divided into five *independent* modules: **Introduction to the Earth System, Geosphere, Hydrosphere, Atmosphere,** and **Cosmosphere**. References to the **biosphere** are woven throughout the workbook. Although this *Explorer's Guide* is closely aligned with and makes frequent reference to the text *Earth Science* by Tarbuck and Lutgens, it could be easily adapted for use with other texts by modifying specific references to page numbers to conform to the textbook you use.

Focus on Instructor Flexibility

In recognition of the great diversity of content, organization, and audience found in introductory Earth science courses, this workbook is designed for maximum flexibility of use. You may pick and choose from a menu of exercises and instructional approaches, using only those that are best suited to your unique teaching style, course format, and student population. *It is not necessary to use all of the modules or to assign all of the questions in any given module.* The workbook includes two types of questions:

1. **Short Answer**, which take approximately 5-30 minutes to complete;
2. **Longer Answer**, which require more time and effort on the part of the student.

Pedagogical Approach

Each module incorporates the following features:

1. **Recurring Themes**

   The following themes are discussed in Module 1 and are threaded throughout the workbook:

   a. Interactions of Spheres
   b. Scale
   c. Cycles
   d. Energy
   e. Humans and the Earth System

2. **KWL Chart**: Assesses prior knowledge and interests by asking students to consider what they already know (K) and what they would (W) like to know about a given topic. At the end of the module, students will reflect on what they have learned.

3. **"Earth System Snapshots"**: Short essays that complement and extend the "Earth as a System" boxes in *Earth Science*.

4. **Questions** that require students to apply what they have learned. Diverse types of assignments and skills are employed including writing, plotting and interpreting graphs, synthesis of information from diverse sources, and application of simple mathematical relationships.

5. **Concept Maps**: Graphical organizers that help students to discover relationships among concepts. Module 1 includes an introduction to concept mapping and gives step-by-step instructions.

6. **WebQuest**: An Internet-based exercise that includes a task that is doable and interesting, links to Web resources needed to accomplish it, and a description of the desired learning process.

7. **Summing Up the Sphere**: Using information from the "Earth System Snapshots," the "Earth as a System" boxes in *Earth Science*, and their own ideas, students describe interactions of each sphere with Earth's other components.

8. **KWL Revisited**: Students return to the KWL chart to correct any misconceptions, ask any unanswered questions, and consider what they have learned.

A **Glossary** and list of **References** are found at the back of *An Explorer's Guide*. The glossary provides students with definitions for terms that are not included in *Earth Science*. References are divided by level into introductory and advanced.

Why Incorporate Active Learning?

Ample research shows that students understand and retain concepts better when they process information though writing, discussion, or application than when they passively listen, take notes, and memorize for a test. Yet, there are significant barriers to the implementation of active learning. If you have a large class, you may lack the time to grade additional assignments. Introductory Earth science courses typically cover a multitude of topics in a single semester, so you may feel that you cannot afford to introduce new topics or set aside class time for students to work on questions. *An Explorer's Guide* to the *Earth System* is intended to help you introduce systems thinking and active learning to your course without taking an "all or nothing" approach. Some portions of the workbook can be used to punctuate a traditional lecture with short, ungraded exercises that require students to summarize, discuss, or apply what they have learned. Longer assignments can be completed out of class as homework or extra credit and may be graded or ungraded. If you have a large class, you may choose to discuss, but not grade, longer assignments that will then be used as the basis for exam questions. The accompanying **Teaching Guide and Solutions Manual** includes additional teaching tips, a matrix that correlates key concepts with suggested assignments, scoring rubrics to help streamline grading, and multiple-choice, true-false, and fill-in-the-blank questions designed to measure student understanding of selected questions from the workbook.

Module 1
Introduction to the Earth System

When we try to pick something out by itself, we find it attached to everything else in the universe.
— John Muir

We shall not cease from exploration. And the end of our exploring will be to arrive where we started from and know the place for the first time.
— T.S. Eliot

Introduction to This Workbook

The purpose of *An Explorer's Guide to the Earth System* is to engage you in active inquiry about our home planet. Through a series of readings, questions, and activities, you'll gain an appreciation for the interconnectedness of Earth's rocks, water, air, and living things.

You're studying Earth science at a particularly exciting time! New technology and ever-expanding computer power are allowing us to see Earth from above and to model complex processes and interactions. As you read this, thousands of satellites are circling Earth, collecting images of and data about its volcanoes, vegetation, cloud patterns, ocean temperatures, and air quality. The Space Shuttle and International Space Station have allowed astronauts to take countless stunning photos of our planet from the unique vantage point of space (Fig. 1.1).

You can use the Internet to retrieve footage of erupting volcanoes, track today's global earthquake activity, and access satellite images of current weather systems. These global views make it possible for us to understand Earth in ways that scientists could once only dream of. We have come to look at Earth as a single system with complexly interacting parts. This workbook is designed to help you study Earth from the unifying perspective of **Earth system science**.

Why Use a Systems Approach?

The Earth and space sciences have traditionally been subdivided into the disciplines **geology**, **oceanography**, **meteorology**, and **astronomy**. In this approach, the geologist who studies the solid Earth doesn't know about the latest findings of the oceanographer, who may have trouble understanding the jargon used by a meteorologist or an astronomer. This situation of a lone scientist studying only one of Earth's multiple subsystems has changed in concert with our ability to see Earth as a single integrated whole and our growing recognition that global environmental problems such as ozone depletion and global warming can only be understood

Figure 1.1 Photo taken from the Space Shuttle of the International Space Station backdropped by the darkness of space and the cloud-covered Earth at its horizon.
(*Source:* NASA Headquarters.)

through the work of multidisciplinary teams of scientists who also communicate with social scientists and policy makers.

An Explorer's Guide will lead you through a series of investigations designed to show linkages among the processes and events that shape planet Earth. You may be surprised by how a change in one part of the Earth system can produce unexpected responses elsewhere. We are learning that humans are powerful agents of change, and one of the major goals of Earth science is to understand current and past changes in order to predict how human actions will influence Earth's water, air, and organisms in the future. Another motivation for studying Earth science is to find ways to reduce risk from natural hazards such as earthquakes, volcanic eruptions, floods, and severe weather.

As you proceed through this workbook, you'll gain a sense of your place in the Earth system. You'll also acquire problem-solving skills that you can use elsewhere. And last, but not least, you'll hopefully develop a curiosity about planet Earth that will continue after this course is over and lead you to new explorations.

An Explorer's Guide is closely aligned with and makes frequent reference to *Earth Science* by Tarbuck and Lutgens, so you'll want to have that text handy. There are five modules in the workbook: this introductory chapter, and chapters corresponding to the following subsystems or spheres of the Earth system: **geosphere**, **hydrosphere**, **atmosphere**, and **cosmosphere**. There is no separate module devoted to the **biosphere** (Earth's living things), but you'll find discussions of the considerable role of life in the Earth system threaded throughout the other modules.

Your instructor may choose to have you do all or selected parts of a given module and will provide guidance about which portions are best suited to the goals of your Earth science course. He or she may not grade all assignments turned in for credit, but may decide instead to include questions on exams that measure your understanding of assigned essays and questions. Your professor may elect to provide you with a *rubric* for selected assignments. A rubric is a rating scale and list of criteria to be used in assessing your work. Give the rubric careful consideration before you begin your work so that it can guide your successful completion of the assignment.

Part I.   Tools Used in This Guide

A. Themes

All explorers need guideposts for their journey, and ours will be the following themes that will be repeated throughout *An Explorer's Guide to the Earth System*. Look for this icon as your sign that one or more of the guiding themes described below is being revisited.

1. **Interactions of Spheres**

    Earth is made of several subsystems or spheres that interact to form a complex and continuously changing whole called the **Earth system**. This workbook is organized into five modules: an introductory chapter and modules devoted to the **geosphere** (solid Earth), **hydrosphere** (water), **atmosphere** (air), and **cosmosphere** (the universe surrounding our planet). Study of the **biosphere** (living things) cuts across all modules.

2. **Scale**

    Processes operating in the Earth system take place on spatial scales varying from fractions of millimeters to thousands of kilometers, and on time scales that range from milliseconds to billions of years.

3. **Cycles**

    The Earth system is characterized by numerous overlapping cycles in which matter is recycled over and over again.

4. **Energy**

    The Earth system is powered by energy from two major sources: the Sun and the planet's internal heat.

5. **Humans**

    People are part of the Earth system and they impact and are impacted by its materials and processes.

B. KWL Chart

Sometimes it's helpful to think for a moment about what you already know before beginning a new investigation of a topic. At the start of each module, you will be asked to fill out a chart like the one shown in Table 1.1.

The parts of the chart are: K: What I know; W: What I want to know; and L: What I learned. This exercise will give you a chance to assess your prior knowledge of the subject and to identify things you are curious about. You'll return to the KWL chart at the end of the module to correct any misconceptions you may have had, ask any unanswered questions, and reflect upon what you've learned.

Go ahead and fill out Table 1.1 with your existing knowledge and questions about planet Earth. You will come back to this chart at the end of the course to see what you've accomplished, check if your questions have been answered, and ask any new questions arising from your investigations in this workbook.

TABLE 1.1   KWL chart about planet Earth.

K	W	L
What I Know	What I Want to Know	What I Learned

C. Earth System Snapshots

Many of the events and processes that shape Earth are very complex and take place over millions of years, so it's hard for us to observe them directly. Each module features several essays called "Earth System Snapshots," which capture some small portion of the Earth system in order to highlight interactions among its spheres. For example, in the Geosphere Module, an essay entitled "Lahar!" describes how water (hydrosphere) combines with volcanic ash (geosphere) to form one of the most dangerous hazards associated with volcanic eruptions.

D. WebQuests

The Internet is a treasure trove of Earth science data, illustrations and photos, maps, and movies. Use of the World Wide Web can enrich your study of the Earth system by providing such visual and written information as satellite images, accounts of the most recent volcanic disaster, animations of ocean currents, and videos of landslides, floods, and hurricanes. But these things can be difficult to find. You've probably had the experience of feeling overwhelmed while searching the Internet. The huge amount of information and variable quality of that information can lead to squandered time while doing open-ended searches. Each module in this workbook has a type of *guided*, inquiry-based Internet exercise known as a *WebQuest*. This tool, developed by Bernie Dodge and Tom Marsh at San Diego State University, consists of a doable task, learning advice, and links to Web sites selected to facilitate the task. A WebQuest requires you to synthesize and apply information from several sources to answer a question or solve a problem.

E. Glossary

Key terms shown in bold are included in the glossary at the back of *Earth Science* by Tarbuck and Lutgens. Bolded terms that are not listed in that glossary are defined at the back of this *Explorer's Guide*.

F. References

At the back of this workbook you'll find a list of references that will help you continue your explorations of the Earth system.

G. Concept Maps

You've probably heard the phrase "a picture is worth a thousand words" countless times. Concept maps are *visual* representations of a topic. You can think of concept maps as a tool that allows you to "show what you know." You'll use them in your exploration of the Earth system to test your understanding of key terms and ideas.

Concept maps are a powerful tool because they help you to make sense of and remember information. Think of the last time you studied for a test. Did you cram the night before, trying to stuff all the facts into your head only to soon forget most or all of them? (But hopefully not until *after* the exam.)

1. To get a sense of how organizing information can help you recall it, look at the words listed below for 20 seconds and attempt to memorize them. Write down what you can recall or have someone quiz you.

Apple	Table	Snail	Banana
Sky	Green	Furry	Short
Motorcycle	Plum	Yellow	Purple

2. Now look at the list below and do the same thing.

Orange	Blue	Dog	Tall
Banana	White	Cat	Narrow
Grape	Red	Bird	Rough

3. Which list was easier to memorize? Why?

Chances are, you had better luck with the second list because the words are arranged in groups of related things. You may have also been helped by a previous association in your mind of the American colors red, white, and blue. Describing how things or ideas are related is what concept mapping is all about.

4. Here are the steps to follow in making a concept map:
   a. Make a list of the major terms or concepts you know about the topic. Concept words are typically nouns or adjectives and are placed in circles. When you are starting out, it may be helpful to jot each concept on a Post-it™ note or index card for ease in moving them around until you are satisfied with the organization of your map.
   b. Arrange the concepts with the most general at the top and more specific concepts below them.

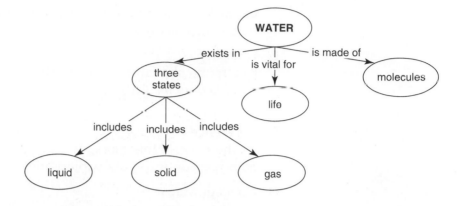

Figure 1.2 Concept map about water.

c. Draw lines between related concepts. Label the lines to accurately describe the relationship between the linked concepts. Examples of linking words:

Includes/Including	Leads to	Produces
Such as	Causes	Is made of
Comes from	Is measured by	Contains
Is based on	Has/Have	Is/Are
May indicate	Can be	Impacts

The concept map about water shown in Figure 1.2 provides a simple example of linked concepts placed in hierarchical order with the most general concept at the top.

5. Pointers for making a concept map
   - In general, concepts and links should contain no more than 2–3 words. Avoid using definitions as concepts.
   - Don't be frustrated by your first attempts. Concept maps take practice; making and using them will become easier with experience.
   - Don't expect your map to look exactly like your classmates'. There is more than one way to accurately show relationships among related concepts. Remember that your map shows *your* understanding.
6. Give it a try. Fill in the question marks on the concept map shown in Figure 1.3.

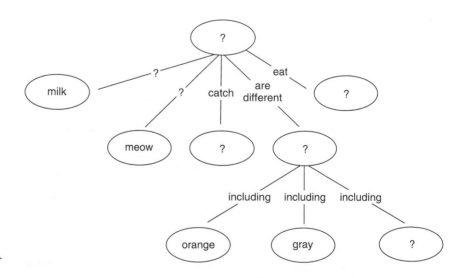

Figure 1.3 Concept map exercise.

7. Now, try one of your own. Let's start with something familiar. Using the steps outlined above, make a concept map that includes the following words:

Apple	Salty	Ice cream	M&Ms
Sweet	Potato chips	Chocolate	Healthy

Add at least three concepts of your own and show the appropriate links with other concepts. Use your imagination to see what you can come up with. You'll have a chance to compare your work to that of others during a class discussion.

H. Summing Up the Sphere

Each module concludes by asking you to synthesize the "Earth System Snapshots," "Earth as a System" boxes in your text, and your own ideas to describe interactions of that module's sphere with Earth's other components.

Part II. Questions

A. Themes
1. Systems
a. *What is a system?*

As a first step in understanding why we will use an Earth system perspective to explore our planet, let's look more closely at the concept of a system. Because the term is commonly used in everyday life, you probably already have some ideas.

1. Write your own definition of "system" in the space below.

2. Compare your definition with that of the person next to you. Did your definitions have anything in common? How did they differ?

3. Review the definition of a system as given on p. 9 of *Earth Science*. Can you add anything to your definition?

4. Make a list of three systems you already know about and identify the parts that make up each.

b. *What are some important aspects of systems?*

1. Systems can be open or closed. An **open system** is characterized by the exchange of matter and energy with its surroundings. For example, all living things are open systems because they take in food, reject wastes, and exchange energy with the

environment. A **closed system** can exchange energy, but not matter, with its surroundings. An unopened package of Oreos™ is an example because no cookies will be added from the surroundings and none will be taken from the system (until you open it and get started, that is).

2. Is Earth an open or closed system?
Explain your reasoning.

3. A car is a good example of an open system.
   - What are the parts of the car system?

   - What is the role of each part in the operation of the system?

   - What happens to the system if one of these parts is removed?

   - What are the *inputs* of energy and matter from the environment to the car system?

   - What are the system's *outputs* to the environment?

c. *Feedback* (Note: This expands upon the discussion of feedback found on p. 9 in *Earth Science*.)
Because the components of a system are complexly linked, a change in the system may led to further changes through a phenomenon known as **feedback**. Feedback is a word with which you're already familiar. When your professor provides feedback on a draft of your term paper, he or she supplies information that you can use to change the paper and hopefully get a better grade. Thus, the feedback is input back into the system (your writing) to guide further changes to the system. As applied to the Earth system, **positive feedback** serves to amplify the effects of the initial change, while **negative feedback** acts to dampen them.

Positive feedback might be called a "vicious circle." A familiar example is exponential population growth. An increase in human births leads to an increase in the human population, which leads to more births, and so on.

In contrast, negative feedback offsets the effects of an initial change and tends to balance a system. The classic example is a thermostat attached to a furnace. You set the thermostat to a desired temperature. If the air in the room is below the target temperature,

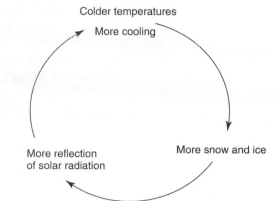

Figure 1.4   An example of positive feedback in the Earth's climate system.

the thermostat signals the furnace to turn on, sending warm air to the room. If the air in the room is above the desired temperature, the thermostat causes the furnace to shut off. If the outside temperature drops, the room will get cooler, causing the furnace to turn on, and so on. The thermostat serves to regulate the temperature in the room so that you are comfortable.

Here's an example of positive feedback in the Earth system: If some change in the Earth system, such as blocking of solar radiation by volcanic dust or a decrease in the energy output of the Sun causes global cooling, more ice and snow will form. Because ice and snow are highly reflective, less solar energy is absorbed at Earth's surface, leading to more cooling, which leads to more ice and snow, and so on. This example of **positive feedback** is shown in Figure 1.4.

Clouds provide an example of **negative feedback** as illustrated in Figure 1.5. Some initial change in the Earth system causes warming, such as an increase in carbon dioxide in the atmosphere. Increased warming leads to increased evaporation, forming more clouds. The clouds reflect incoming solar radiation, leading to less warming. The initial warming is *counterbalanced* by the formation of clouds, which cause cooling. However, as is the case with many processes in the very complex Earth system, it's probably not so simple. The role of clouds is one of the big uncertainties facing scientists studying climate change. Some clouds

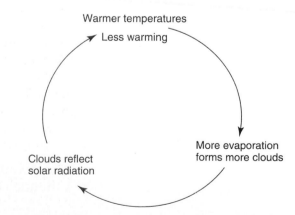

Figure 1.5   An example of negative feedback in the Earth's climate system.

may lead to warming instead of cooling by trapping some of the infrared radiation emitted by the Earth.

Test your understanding of feedback with the following problems. Circle your choice in each pair of words.

1. **Don't change that dial . . .**
   a. A popular radio station decides to sell more ad time. This causes the station's revenue to (increase/decrease).
   b. The increased ratio of ads to music causes irked listeners to switch radio stations. Fewer listeners means that advertisers are less willing to buy ad time. This (reinforces/counterbalances) the initial increase in the station's revenues.
   c. This is an example of (positive/negative) feedback.

2. **Gazelle herd**
   a. An increase in the birthrate in an gazelle herd (increases/decreases) the gazelle population.
   b. An increase in the gazelle population (reinforces/counterbalances) the initial increase in birthrate.
   c. This is an example of (positive/negative) feedback.

3. **Sun and snow**
   a. Ice and snow reflect more solar radiation than does open water. An increase in global temperatures (increases/decreases) the area covered by ice and snow.
   b. Melting of ice and snow causes less solar radiation to be reflected (and more to be absorbed) at Earth's surface. This (reinforces/counterbalances) the initial increase in temperature.
   c. This is an example of (positive/negative) feedback.

2. **Earth's Spheres**

What are the parts of the Earth system? Read pp. 5–8 in *Earth Science* and then write a brief description of each of the Earth's interacting subsystems or "spheres."

3. **Scale**
   a. Match each process or event with the letter that most accurately represents its place on Figure 1.6. Don't worry if you feel unsure of your answers. The goal is to get you thinking.
      • Deep ocean circulation
      • Origin of Earth
      • The development of a soil
      • A volcanic eruption
      • Mineral and fossil fuel development
   b. How is the nonrenewable nature of mineral resources and fossil fuels related to the time scales of the processes that form them?

   c. Where would a meteorite impact plot on the diagram? A thunderstorm?

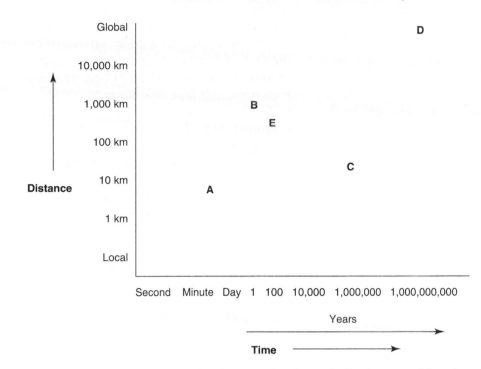

Figure 1.6   Characteristic spatial and temporal scales in the Earth system. Note that the scales of the axes are not linear. (Modified from *Earth System Science; A Closer View*, NASA Headquarters.)

4. **Cycles**
   The Earth system is characterized by numerous overlapping cycles in which matter is recycled over and over again.
   a. Briefly describe at least two cycles that affect planet Earth and its inhabitants.

   b. What is the time scale of each of the cycles you have listed?

   c. The hydrologic cycle
      1. Study the diagram of the hydrologic cycle shown on p. 10 in *Earth Science*. What are the storage places or **reservoirs** for water in the Earth system?

         Living things are another reservoir for water (*you* are an example — 65% of your body is water). Plants absorb water and return it to the atmosphere in a process called **transpiration**.
      2. Write the autobiography of a water molecule that has "been around" (the hydrologic cycle, that is). Your molecule must have spent time in each of Earth's major spheres. Make sure to mention the processes by which the molecule makes its way from one reservoir to another.

5. **Energy**

The Earth system is mostly powered by energy from two sources: the Sun and internal heat. Indicate the primary source of energy for each of the following events and processes:
- Evaporation from the sea surface
- Volcanic eruption
- **Photosynthesis**
- **Geyser**
- Earthquake

6. **Humans and the Earth System**

Humans are part of the Earth system and they impact and are impacted by it. Give two examples of how people are changing the natural hydrologic cycle.

B. **Earth Reporter**

As you begin this course, you will also begin a log of Earth activity. Each week, you'll use newspaper reports and the Internet to prepare an account of the size and location of current events in the Earth system. This activity will help you get a sense of the geographic framework of Earth processes such as earthquakes, volcanoes, hurricanes, and tornadoes and of the frequency with which they occur over a period of several weeks. Your instructor will provide a format for reporting your data. These Web sites will give you a good starting place.
- National Earthquake Information Center:
  http://neic.usgs.gov/
- Volcano World:
  http://volcano.und.edu/
- Earthweek – A Diary of A Planet:
  http://www.earthweek.com/
  (Refer to "This week's Earthweek in Adobe Acrobat PDF format.")

WebQuest - Earth: The Just-Right Planet

The Task

According to the Terran Tourism Board (TTB), Earth is losing tourist dollars to its neighboring planets. The members of the Board have asked you to design a brochure entitled "Earth: The Just-Right Planet." The TTB is expecting you to highlight at least four ways in which Earth is unique among the **terrestrial planets**. You will be paid a commission for every new tourist you bring to Earth, so make sure that your brochure is persuasive!

The Process

First, you will need to do some research. You probably already have some ideas about why Earth is a special place, but in order to make your case, you'll need to know more about the other terrestrial planets and their "spheres" (hydrosphere, geosphere, biosphere, atmosphere). Use the Web sites below to summarize your findings, and then prepare a roughly one-page brochure that will convince extraterrestrials that our planet is the one to visit.

Resources

- Welcome to the Planets:
  http://pds.jpl.nasa.gov/planets/
- Exploring the Planets:
  http://www.nasm.edu/ceps/etp/
- The Nine Planets:
  http://seds.lpl.arizona.edu/nineplanets/nineplanets/nineplanets.html
- Views of the Solar System:
  http://www.solarviews.com/eng/homepage.htm
- Windows to the Universe:
  http://www.windows.umich.edu/

Conclusion

So, did the TTB hire the right person for the job?

Module 2

Geosphere

The Geosphere

The **geosphere** is the solid part of the Earth system, including the rocks below your feet. The outermost part of the solid Earth, the **lithosphere**, is broken into numerous tectonic plates. Unequal distribution of heat within Earth generates plate motion and leads to mountain building, earthquakes, and volcanic eruptions.

In the first part of this module, you'll explore the materials of which the geosphere is made. In the second part, you'll investigate the dynamics of the solid Earth including plate tectonics, earthquakes, and volcanic eruptions. Along the way, you may be surprised by how changes in the "rock solid" part of our planet influence you and your environment.

This module contains several readings, questions, and problems to assist you in your study of minerals, rocks, and soils and the processes that affect them. Your instructor may choose to have you do all or selected parts of this module.

Geosphere 1: Earth Materials

Learning Objectives

After completing the following assignments, you should be able to answer these questions:

1. What is the formal definition of a mineral?
2. What properties can be used to identify minerals?
3. What processes form igneous, metamorphic, and sedimentary rocks?
4. How does the rock cycle work?
5. What factors influence the formation of soil?
6. What kinds of soil are found in arid, humid, and tropical climates?
7. How can rock and soil types be used to interpret Earth history?

Part I. Getting Started: KWL Chart

Before you begin your study of Earth materials (minerals, rocks, soils), take a few moments to fill out the chart below. You will come back to it at the end of this module to see what you've accomplished, check if your questions have been answered, and ask any new questions arising from your investigations.

TABLE 2.1 KWL chart about Earth materials.

K What I Know	W What I Want to Know	L What I Learned

Part II. Earth System Snapshots

Sinkholes, Stalactites, and Stone Monuments

Themes: Interactions of Spheres; Humans and the Earth System

You've probably learned in your Earth science class that a surefire way to identify the mineral *calcite* (or the rocks *marble* and *limestone*, which are made of it) is with a drop of hydrochloric acid. The calcite will fizz as it dissolves, releasing bubbles of carbon dioxide 1$CO_2$2. This behavior is important for a number of reasons (in addition to your rock and mineral quiz). Limestone and marble underlie many parts of the world, including portions of New Mexico, Kentucky, and Florida. Due to the solubility of carbonate rocks in water, such areas often have spectacular caves such as the Carlsbad Caverns in New Mexico and Mammoth Cave in Kentucky. Even unpolluted rainwater is acidic because it dissolves some CO_2 as it moves through the atmosphere, forming weak *carbonic acid* (see pp. 70–71 in *Earth Science*). This rainwater percolates downward to form groundwater, which can serve as both an agent of erosion to form caves and sinkholes and as an agent of deposition, when calcium carbonate is re-precipitated to form **stalactites** and **stalagmites**. In 1981, the people of Winter Park, Florida, witnessed a spectacular example of groundwater erosion when a rapidly forming sinkhole swallowed a house, part of a swimming pool, and several cars. **Karst** topography is the name given to landscapes that have been shaped by the dissolving action of groundwater. You can see some examples in Figure 4.34 on p. 125 in *Earth Science*.

Figure 2.1 Effects of acid rain on building materials. Compare sharp edges on undamaged areas (A) with rounded edges and rough surfaces (B) produced by acid rain. (*Source*: Venessa Miles/Environmental Images.)

The solubility of carbonates also affects their use as building stones and monuments. Headstones made of limestone or marble don't last as long in humid climates as do those made of granite. Limestone and marble are particularly vulnerable in areas that are affected by acid rain, such as the northeastern United States. For example, the Jefferson Memorial and other buildings in our nation's capital are deteriorating due to acid precipitation (Figure 2.1), as are such historical treasures as the Taj Majal in India, the Acropolis in Greece, Westminster Abbey in London, and the ancient Mayan ruins of Mexico.

Part III. Questions

A. Short Answer

1. Is ice a mineral? Why or why not?
   (Hint: Review the definition of a mineral on p. 21 in *Earth Science*.)

2. How does the formation of rock salt reflect interaction of the geosphere with other Earth spheres?

3. Would you expect to find limestone on the Moon? Why or why not?

4. Refer to the rock cycle diagram on p. 41 of your textbook.
   a. Identify the Earth spheres that are involved in each of the following processes:
      - Cementation
      - Weathering
      - Transportation
   b. Identify the source(s) of energy that drive each of the following processes:
      - Melting
      - Weathering
      - Metamorphism
      - Transportation
5. Describe three ways that the rock cycle would differ if there were no water on Earth.

B. Longer Answer
1. Soil Sleuth

As a huge fan of tennis, you are elated when you win a trip to watch the Wimbledon Championships! But your excitement about tennis is soon spoiled by another type of excitement when one of the players has her best tennis racket stolen. She is terribly distraught, insisting that the police must find her racket right away, for she cannot possibly compete without it. The crowds complicate investigation of the theft, but the police have managed to narrow it to three suspects, all of whom had motive, means, and opportunity: Alberto, Alice, and Allistair. Each was heard to have a heated argument with the victim of the theft (who isn't very easy to get along with) at a reception held the previous afternoon. It's possible that the racket was stolen as an act of revenge. Fortunately, there is another clue: the thief got careless and left some soil from his or her tennis shoe at the scene. Can the soil help nab the culprit? This is your chance to apply your knowledge of where different types of soil are found. Refer to pp. 80–81 and Appendix C in *Earth Science*. It may be helpful to know that Alberto is from Manaus, Brazil, Alice hales from Lincoln, Nebraska, and Allistair is from near Wiluna, Australia.

A lab report describes the sample as brown with minute fragments of grass. It does not react in dilute hydrochloric acid. A chemical analysis reveals that the soil is rich in iron and aluminum and contains **humus**.

- Who did it?
- Which soil-forming factor (pp. 77–79 in your text) did you rely upon to identify the perpetrator of the crime?

Of course, in the real world, your detecting skills would be much more greatly challenged because there are several variables that control soil formation. Read pp. 77–79 in *Earth Science* and then write a brief summary of each of the five soil-forming factors.

2. **Written in Stone**
   As discussed on pp. 192–197 of your text, Alfred Wegener had several lines of evidence for his hypothesis of continental drift.
   a. Briefly summarize the types of evidence.

   b. One of the items on your list should be ancient climate. This is where the rock record comes in. Because interactions among the geosphere, atmosphere, hydrosphere, and biosphere produce rocks, minerals, and soils, these materials provide a record of the environment in which they were formed. Think of sequences of layered rocks like those so beautifully exposed in the Grand Canyon (p. 289 and 292 in your text) as a kind of "storybook" that reveals Earth's past. Rock layers help to reconstruct the movement of continents through time because one of several controls on an area's climate is its latitude. For example, in the modern world, we see that tropical climates are found nearer to the equator than are areas covered with ice sheets (see Figure 2.2 and pp. 524–528 in your text). Another vital clue to the climate of a place is the types of organisms living there, and fossils are invaluable for reconstructing past environments.

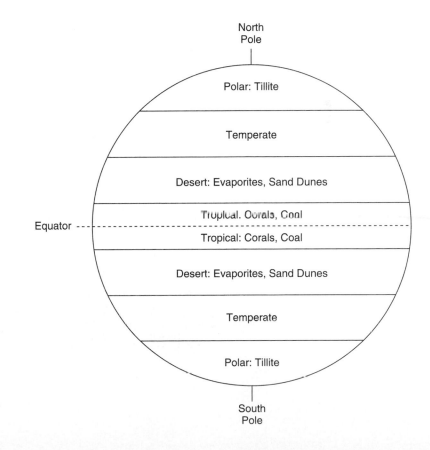

Figure 2.2 Generalized pattern of Earth's present-day climate zones and associated Earth materials.

TABLE 2.2 Climate clues in Earth materials.

Rock or soil type	Climate clues including **temperature** (hot/warm, cold/cool), **moisture** (wet or arid). Also indicate whether the rock or soil forms in a **continental** or **marine** environment.
• Coal	
• Glacial till	
• Rock salt	
• Laterite	
• Bauxite	
• Rock gypsum	
• Limestone with corals (Hint: corals like warm water.)	
• Sandstone with dunes and lizard tracks	
• Shale	

c. Your textbook describes two rock types used by Wegener as evidence for continental drift. Find them in Table 2.2 and describe the climatic clues they provide. Use your text to complete the table. Be as thorough as possible.

d. Before you can use rock sequences to interpret changes in the environment through time, you need to understand an important rule known as the law of **superposition**. Read about this law on p. 289 in your book and write a description in your own words.

Questions e-j refer to Figure 2.3.

e. Based on what you have learned, what is the oldest rock or soil layer in column A?
   In column B?
   In column C?

f. What is the youngest rock or soil at location A?

   How do you know?

   What event formed this rock?

g. Which continent, A, B, or C has probably moved closer to the equator through time?
   Explain.

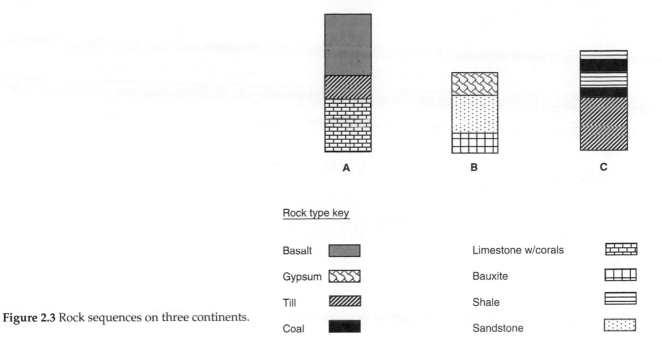

Rock type key

Basalt

Gypsum

Till

Coal

Limestone w/corals

Bauxite

Shale

Sandstone

Figure 2.3 Rock sequences on three continents.

h. Which has moved through time from a lower to a higher latitude? How do you know?

i. Which has experienced a change from a very wet to an arid climate?
How do you know?

j. Look at p. 195 in *Earth Science*. How did the position of India change during the time interval between 200 million years and 50 million years ago?

Which of the rock columns best matches the rock sequence you would expect to find on India?
Explain.

Part IV. Concept Map About Minerals

Fill in the question marks in the concept map shown in Figure 2.4. Add three new concepts and show how they link with other concepts.

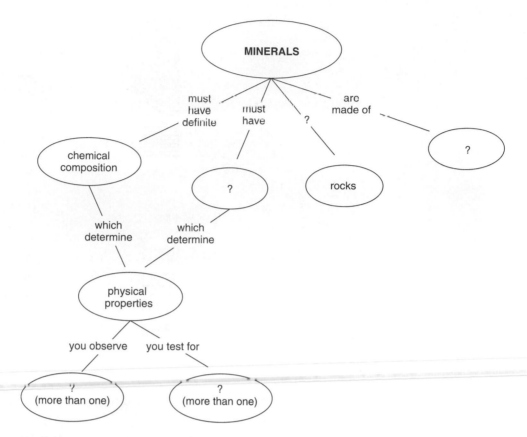

Figure 2.4 Concept map about minerals.

Geosphere 2: Earth Dynamics

Learning Objectives

After completing the following assignments, you should be able to answer these questions:

1. What is a **tsunami**?
2. How do tsunamis compare to wind-driven ocean waves?
3. How fast are tsunamis? How do we know?
4. What roles do plate tectonics and chemical weathering play in the **carbon cycle** and climate change?
5. How does plate tectonics affect the biosphere?
6. What are some examples of how volcanoes affect Earth's hydrosphere, atmosphere, and biosphere?

Part I. Getting Started: KWL Chart

Before you begin your study of Earth dynamics (plate tectonics, earthquakes, and volcanoes) take a few moments to fill in Table 2.3. You will come back to it at the end of this module to see what you've accomplished, check if your questions have been answered, and ask any new questions arising from your investigations.

TABLE 2.3 KWL chart about Earth dynamics.

K What I Know	W What I Want to Know	L What I Learned

Part II. Earth System Snapshots

Continental Drift and Kangaroos

Themes: Interactions of Spheres

Imagine yourself living in India at the time of Pangaea (see Figure 7.7 on p. 198 in *Earth Science*). It's currently quite cold there, but you're a passenger on a lithospheric plate that's heading for the equator. In a few million years, you'll need to pack away your parka and assemble a summer wardrobe. In other words, you'll need to adapt to your new environment (there's no need to hurry, though, because you're only moving about as fast as your fingernails grow). As illustrated by the warming of India's climate as it drifted northwards, plate motion has a profound effect on the physical environment. And, due to the interconnectedness of Earth's spheres, changes in the environment must inevitably affect living things and their evolution.

Plate tectonics affects the biosphere in many direct and indirect ways:
- The joining of continents merges previously separated populations of terrestrial creatures. For example, the emergence of the Central American land bridge about three million years ago led to the exchange of organisms between North and South America. Many of the South American marsupials that had thrived in isolation became extinct when placed in competition with placental mammals from North America.

- Break-up of continents creates populations that may evolve differently. For example, the long-term isolation of Australia from other continents has led to the development of a unique and diverse fauna that includes kangaroos, wombats, koalas, and platypuses.
- Climate changes when continents drift to different latitudes, and when reconfiguration of continents alters ocean circulation patterns.
- The opening and closing of oceans disrupts marine habitats.
- Volcanism related to plate interaction releases dust and gases which may produce warmer climates due to the **greenhouse effect** or cooler climates due to blocking of solar radiation. Some scientists have proposed that it was volcanism, not a meteorite impact, that killed off the dinosaurs. Large outpourings of lava have been implicated in several **mass extinctions** through geologic time.
- Plate tectonics causes uplift, producing mountains that change atmospheric circulation and create rain shadow deserts.

Lahar!

Themes: Scale; Humans and the Earth System

What picture comes to mind when you think of a volcanic eruption? If you are like most people, your mental picture may feature glowing fountains of lava or huge billowing clouds of ash. But **lahars** (p. 90 and 238 in *Earth Science*), or volcanic mudflows, although a less flashy product of volcanism, can be just as lethal. Lahars form when unstable layers of ash and other volcanic debris become saturated with water and flow down slopes like fast-flowing concrete, usually following stream channels (Figure 2.5). The water can come from rain, melted snow and ice, or lakes.

The deadly power of volcanic mudflows was tragically demonstrated in 1985 when 25,000 lives were lost due to lahars triggered by a small eruption of the volcano Nevado de Ruiz in Colombia, South America. Nevado means "snowy" in Spanish and, despite the fact that Nevado del Ruiz is located near the equator, its high elevation (17,453 feet; 5,321 m) results in a year-round cap of ice and snow. The ice and snow melted in contact with hot **pyroclastic** material, sending torrents of hot ash, debris, and mud down three major river valleys on the volcano's sides. The lahars reached maximum velocities of over 60 km per hour and buried several villages in their path. The worst hit was Armero, located 48 km from the volcano, where 23,000 people were killed.

Although the Nevado del Ruiz lahars were the single most catastrophic in recorded history, in terms of cumulative deaths, lahars are in some ways as dangerous as **pyroclastic flows**. They rush down stream valleys at speeds that may exceed 100 km per hour and may travel hundreds of km from their source. They are the only volcanic hazard that doesn't require an actual eruption. Heavy precipitation may trigger a lahar long after the eruption is past, catching people when they believe that the danger is over.

One of the reasons that Mt. Rainier in the Cascades of Washington state is considered by many to be America's most dangerous volcano is that it, like Nevado del Ruiz, is covered with snow and ice and has steep sides. Adding to the risk is the fact that 100,000 people live in river valleys around Rainer and many of their homes are built on lahars that flowed down the volcano hundreds to thousands of years ago. The next eruption is likely to launch lahars that take similar paths. The United States Geological

Figure 2.5 One of 1,300 houses damaged by a lahar triggered by heavy rains on the slopes of Unzen Volcano, Japan, in 1992–1993. (*Source*: U.S. Geological Survey.)

Survey is working to develop an automatic warning system that uses a seismic network to detect vibrations caused by lahars. But a warning system can only reduce, not eliminate, the risk because the travel time of a lahar to the nearest densely populated area may be as little as 30 minutes.

Volcanoes: Natural Sources of Air Pollution

Themes: Interactions of Spheres; Scale; Humans and the Earth System

When someone mentions air pollution, you probably think of cars sitting on a freeway, belching noxious fumes, or of smokestacks pouring waste into the atmosphere. Although human activities are certainly one cause of deteriorating air quality, natural Earth processes also play a role. In addition to lava, volcanoes erupt ash and gases (Figure 2.6; pp. 230–231 in *Earth Science*). The composition of volcanic gases varies, but water vapor is usually dominant, followed in abundance by carbon dioxide and compounds of sulfur and nitrogen with lesser amounts of chlorine, hydrogen, and argon. When volcanic gases and dust enter the atmosphere, they impact the environment on scales that vary from local to global.

Sulfur Dioxide

In the atmosphere, sulfur dioxide $1SO_22$ reacts chemically with sunlight, oxygen, and dust to form tiny particles called **aerosols**, with sometimes drastic consequences. Volcanic aerosols may produce plant-damaging acid rain and a hazy pollution known as **vog**. Vog is a problem in Hawaii, where *each day*, Kilaeua volcano produces about 2,000 tons of SO_2! The health hazards of vog are not completely known, but people living on the

Figure 2.6 Sampling a volcanic vent at Mt. Baker, Washington.
(*Source*: U.S. Geological Survey.)

Big Island of Hawaii have reported headaches, flu-like symptoms, and breathing problems. Sulfuric acid aerosols may also affect global climate by blocking solar radiation, causing cooling. After the 1991 eruption of Mt. Pinatubo in the Philippines, which injected nearly 20 million tons of SO_2 into the atmosphere, global temperatures dropped by as much as 0.5 degrees C in some areas.

Carbon Dioxide

Because CO_2 is a **greenhouse gas**, volcanic eruptions may cause global temperatures to rise rather than drop. Carbon dioxide is colorless, odorless, and because it is denser than air, tends to collect in low-lying areas. If present at high enough levels, CO_2 can have lethal effects as demonstrated by a tragic occurrence at Lake Nyos in Cameroon, West Africa. Lake Nyos fills the crater of a volcano that has not erupted in historical times. In 1986, a cloud of CO_2-rich gas was released from the lake. It hovered near the ground, traveling down valleys at speeds of 20-50 km per hour to several nearby villages where 1,700 people died from suffocation.

Ash

Volcanic ash blown into the atmosphere also represents a significant environmental hazard. It causes breathing problems, damages plants, and may cause buildings to collapse from its weight (Figure 2.7). Ash is also a hazard to aviation, reducing visibility, damaging flight control systems, and causing jet engines to suddenly fail. The danger is increased by the fact that clouds of ash can travel far from their source and are difficult to distinguish

Figure 2.7 House damaged by a heavy accumulation of ash on its roof during the 1994 eruption of Rabaul Caldera, Papua New Guinea. (*Source*: U.S. Geological Survey.)

by sight or radar from ordinary clouds. Fortunately, to date there have been no crashes as a result of jets flying through ash plumes, but there have been several close calls. In the two decades prior to 1996, more than 60 planes, most of them commercial jets, were damaged by in-flight encounters with volcanic ash that in some instances resulted in power loss to all engines and emergency landings.

Part III. Questions

A. Short Answer
Describe three ways that Earth would be different if there were no plate tectonics.

B. Longer Answer
1. Tectonic Tourists
You have been asked by the President of the United States to advise a group of visitors from another solar system about their landing sites on Earth. Their goal is to experience as many Earth processes as possible and they are on a tight schedule, so they'll only have time to land in two places. The aliens have specified that they most want to witness processes that will help them learn about Earth's internal energy.

- What two places do you recommend?

- Why do you think that these places are the best to give our visitors the most experience with the dynamics of planet Earth?

2. **Catch a Wave**
   You're lying on the beach in Hawaii listening to the radio when your tunes are interrupted by an announcement that there's just been a huge earthquake in Chile — 9.6 on the Richter scale! You feel badly for the people who live there, but otherwise don't give it a second thought, because, after all, you are an ocean away. You've been studying earthquakes in your Earth science class, and realize that an earthquake of this size generates more energy than 32 billion tons of TNT. Even so, you have nothing to worry about because, after all, you are an ocean away (about 10,000 km, to be exact).

   You'd better think again. That ocean is an excellent medium for transferring the energy released by the earthquake to the shores of Hilo via a seismic sea wave or **tsunami** (pp. 177–179 in *Earth Science*). Tsunamis are a special kind of ocean wave. Before you have a look at the unique behavior of tsunamis, you need to understand waves in general.

   a. Go to pp. 381–385 in *Earth Science* to read about waves. In the space below, make a sketch that illustrates the following terms:

   | Crest | Trough | **Wavelength** | **Wave height** |

   b. According to your text, what generates most ocean waves?

   c. What is the ultimate source of energy for such waves?

   d. Use the data in Table 2.4 to make sketches that compare the characteristics of typical ocean waves to those of tsunamis.

TABLE 2.4 Comparison of normal ocean waves with tsunamis.

Type of Wave	Period	Height in Open Ocean	Wavelength
Most ocean waves	5-20 seconds	3 m	40 m - 600 m
Tsunami	3,600 seconds (60 minutes)	0.3 - 0.6 m	Up to 837,000 m (520 miles)

e. If you were on a ship in the middle of the ocean, would you notice when a tsunami passed you by?
Why or why not?

f. What energy source(s) produce tsunamis? (See p. 177 in your text.)

Note: submarine landslides, volcanic eruptions, and meteorite impacts also generate tsunamis.

g. Figure 2.8 shows travel times for tsunamis produced by the 1960 earthquake in Concepción, Chile, and the 1964 earthquake in Valdez, Alaska. Each concentric line radiating outward from the earthquake epicenters represents a one-hour tsunami travel time increment.
You can use these travel time curves to calculate the speeds of the tsunamis generated by the two earthquakes. This equation will come in handy:

$$\text{Velocity} = \text{distance , time}$$

Distance between Concepción, Chile, and the Hawaiian Islands: _____ km

Time it took the tsunami to reach Hawaii: _____ hours

The tsunami's speed was _____ km/hr. *Show your work.

h. Now calculate the speed of the Valdez tsunami.
The speed was _____ km/hr. *Show your work.

i. Which tsunami traveled more quickly?

j. In deep water, the speed of a tsunami is approximated by the following expression:

$$\text{Speed} = 2 \overline{g * d}$$

Where **g** = the acceleration due to gravity 19.8 m/sec^{2}2 and **d** = depth of the ocean. Referring to the map on pp. 338–339 in

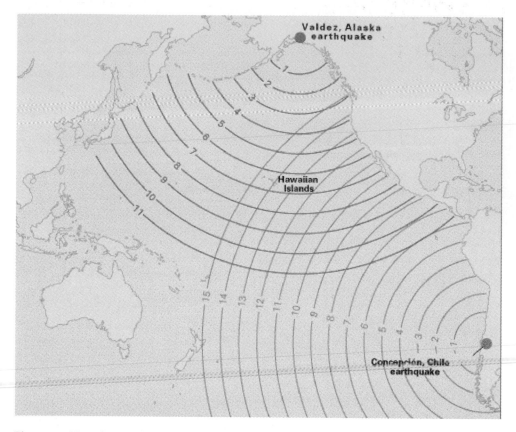

Figure 2.8 Travel times (in hours) for the tsunamis produced by the 1960 Concepción, Chile and 1964 Valdez (Anchorage) Alaska, earthquakes. The Hawaiian Islands are highly vulnerable to tsunamis generated by major earthquakes in the Pacific Ring of Fire (approximate map scale is 1 cm = 1,500 km). (*Source*: U.S. Geological Survey.)

Earth Science and the formula relating tsunami speed to ocean depth, can you explain the different speeds of the Alaskan and Chilean tsunamis?

k. A typical depth in the Pacific Ocean is 4,000 m. Use the equation in the previous question to calculate the speed of a tsunami traveling through water with this depth. *Show your work.

How does your answer compare to the speed you calculated for the Alaskan tsunami using the travel time map?

3. **Volcanoes in the Earth System**
Fill out Table 2.5 using information from your text and the "Earth System Snapshots" on pp. 2-10–2-13 of this workbook. You will also

TABLE 2.5 Effects of volcanism on the Earth system.

Products of Volcanism	Effects on		
	Atmosphere	Hydrosphere	Biosphere
Lava flows			
Lahars			
Pyroclastics			
Gases			

find it useful to check out VolcanoWorld's summary of the effects of volcanism at:

http://volcano.und.edu/vwdocs/frequent_questions/group4_new.html.

4. **Plate Tectonics and the Carbon Cycle**

a. *Background* (Note: This expands upon the discussion of the carbon cycle found in Box 2.1 on p. 52 in *Earth Science*)

The Earth system includes many cycles including the rock and hydrologic cycles. A less familiar but equally important cycle, the **carbon cycle**, describes the flow of carbon and its compounds through Earth's spheres. **Reservoirs** for carbon and carbon compounds include the atmosphere, plants and animals, rocks and soils, and the ocean. A reservoir that takes up carbon from another reservoir is referred to as a *sink*. For example, because carbon dioxide dissolves in seawater, the ocean is a sink for atmospheric carbon dioxide. A reservoir or process that releases carbon to another reservoir is called a *source*. **Deforestation** is a carbon source because burning or decaying wood releases carbon dioxide to the atmosphere.

The carbon cycle can be thought of in terms of two *sub-cycles* operating on different time scales. The **short-term carbon cycle** (Figure 2.9) operates on human time scales and involves interactions between the biosphere, oceans, and soils in the surface and near-surface part of the Earth system. Important processes in the short-term cycle include **photosynthesis** and **respiration**, dissolving and precipitation of carbonate in rocks, sediments, and the shells of organisms, and the exchange of CO_2 between oceans and atmosphere. The short-term carbon cycle will be revisited in

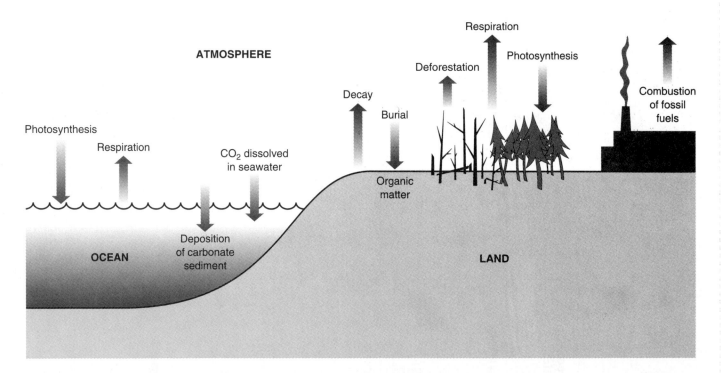

Figure 2.9 Schematic diagram of the short-term carbon cycle. Arrows show the flux of carbon dioxide to (sources) and from (sinks) the atmosphere.

the Hydrosphere Module when you read about oceans as sinks for CO_2 and in the Atmosphere Module when you explore how living things, including humans, affect the flow of carbon in the Earth system.

In contrast, the **long-term carbon cycle** (Figure 2.10) operates on geological time scales (millions of years), and involves two key processes:

- Plate tectonics; and
- Chemical weathering.

The long-term cycling of carbon involves exchange between the **geosphere** and the Earth's other spheres. Sources of atmospheric carbon dioxide include volcanoes and **metamorphism** of carbonate-rich rocks such as limestone. An example is the formation of calcium silicate and carbon dioxide from the reaction of calcite with quartz:

$$CaCO_3 + SiO_2 = CaSiO_3 + CO_2$$

Calcite   Quartz   Calcium silicate   Carbon dioxide

Chemical weathering is a sink for atmospheric CO_2, serving as a kind of "thermostat" that regulates Earth's temperature. As a simple case, consider what happens when a calcium silicate mineral like the one in the metamorphic reaction above is exposed to

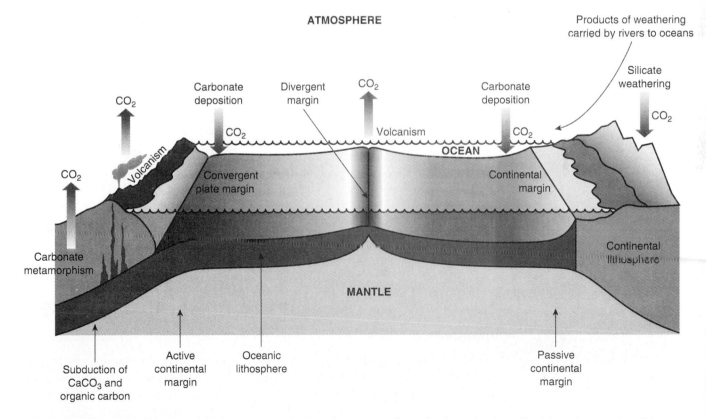

Figure 2.10 Schematic diagram of the long-term carbon cycle. Arrows show the flux of carbon dioxide to (sources) and from (sinks) the atmosphere.

CO$_2$ dissolved in rainwater. There are several steps involved, but the net effect is to *remove* CO$_2$ from the atmosphere and store it in limestone:

$$CaSiO_3 \; + \; CO_2 \qquad = \qquad CaCO_3 \; + \; SiO_2$$

Calcium   Atmospheric   Limestone   Silica dissolved
silicate   carbon dioxide   in rivers and seawater

In summary, plate tectonics and associated mountain building, metamorphism, and volcanism *add*, and chemical weathering of silicates *removes* CO$_2$ from the atmosphere.

b. *Questions*

1. Name two types of plate boundaries where CO$_2$ is released to the atmosphere by volcanism (see Fig. 2.10).

2. Given that an increase of CO$_2$ in the atmosphere increases temperature, how do you think that an increase in the rate of seafloor spreading might affect global climate? Explain.

3. Is the removal of atmospheric CO$_2$ by chemical weathering an example of **positive** or of **negative feedback**? Keep in mind that warming increases the rate of chemical weathering. Explain.

Part IV. Concept Map About Earthquakes

Construct a concept map about earthquakes, using the terms listed below. Add at least three terms of your choosing and show appropriate links to the other concepts.

Faults	P waves
Plate boundaries	S waves
Intensity	Tsunamis
Magnitude	Subduction zone

Part V. WebQuest - Seattle or St. Louis, Eruption or Earthquake?

Introduction

As you have learned in class, plate tectonics provides a framework for understanding the locations of volcanoes and earthquakes. In this exercise, you will explore the geologic settings and histories of two cities to gain an understanding of how geologic hazards might impact them.

The Task

You've just gotten notice that the company you work for is being reorganized and you will be transferred. Because you are an excellent employee, you are given a choice as to where you'd like to go and must choose between Seattle, Washington, and St. Louis, Missouri. As a student of Earth science, you think it might be a good idea to find out which place is at least risk from geologic hazards. You'll use the Web to research the hazards associated with each city and will then synthesize your findings to decide where you'd rather live. You must explain your choice in light of what you've discovered.

Resources

1. General Background
   Plate tectonics. First, you'll need to review plate tectonics and how it works. *This Dynamic Earth,* an on-line booklet from the United States Geological Survey (**http://pubs.usgs.gov/publications/text/dynamic.html**), is a good reference for answering the following questions; you will also find information in Chapter 7 of *Earth Science.*
   - Name and describe the three types of plate boundaries and give an example of each.

   - Refer to the "Plate Tectonics and People" section of *This Dynamic Earth* to write a summary of how plate tectonics impacts people and their property.

   Geologic Hazards

   - United States Geological Survey (USGS) Volcano Hazards Program
     http://volcanoes.usgs.gov/
   - USGS Geologic Hazards Team (Earthquakes/Geomagnetic/Landslides)
     http://geohazards.cr.usgs.gov/
   - Natural Hazards Data from NOAA's National Geophysical Data Center
     http://www.ngdc.noaa.gov/seg/hazard/hazards.shtml
2. For information about St. Louis:
   - St. Louis University Earthquake Center
     http://www.eas.slu.edu/Earthquake_Center/
   - The Mississippi Valley—"Whole Lotta Shakin' Goin' On"
     http://quake.wr.usgs.gov/QUAKES/FactSheets/NewMadrid/
   - Center for Earthquake Research and Information, University of Memphis
     http://www.ceri.memphis.edu/

3. For information about Seattle:
   - USGS Cascades Volcano Observatory
     http://vulcan.wr.usgs.gov/home.html
   - Pacific Northwest Earthquake information
     http://www.geophys.washington.edu/SEIS/PNSN/
   - Mt. Rainier—Living with Perilous Beauty
     http://vulcan.wr.usgs.gov/Volcanoes/Rainier/Publications/FS065-97/framework.html

The Process

In order to complete the task of deciding where to move based on the likelihood of natural disasters, you need to acquire the following information about each city:

1. What is its tectonic setting?

2. Describe its geologic history; has it experienced volcanism and/or seismic activity? If so, what were the sizes and dates of the events? Is it prone to other types of hazards?

3. What are the estimated risks associated with potential geologic activity and how is the risk determined?

4. How well prepared is the city for an earthquake, volcanic eruption, or other hazard?

Learning Advice

- Did you use information from multiple sources?
- Did you crosscheck the information that you found?
- Did you compare information from different sources?

Conclusion

Which city is safer in terms of natural hazards, St. Louis or Seattle? Or, does each have its dangers? Which city will be your new home?

Summary Questions (Geosphere 1 and 2)

Summing Up the Sphere

Using information from the "Earth System Snapshots," "Earth as a System" boxes in your text, and your own ideas, describe interactions of the geosphere with Earth's other components. Give at least two examples of each of the following interactions:

Geosphere-Atmosphere:

Geosphere-Biosphere:

Geosphere-Hydrosphere:

You may also think of examples that involve more than two spheres. Example: Geosphere-Atmosphere-Hydrosphere: Carbon dioxide in the atmosphere combines with rainwater to form carbonic acid, which plays a role in the chemical weathering of rocks.
Your examples:

KWL Revisited

Go back to the KWL charts you filled in for *Earth materials* and *Earth dynamics*. Did you have any initial misconceptions about the topics you've explored? Go back to correct them. Do you have any unanswered questions?

Module 3

Hydrosphere

Suddenly from behind the rim of the moon, in long, slow-motion moments of immense majesty, there emerges a sparkling blue and white jewel, a light, delicate sky-blue sphere laced with slowly swirling veils of white, rising gradually like a small pearl in a thick sea of black mystery. It takes more than a moment to fully realize this is Earth ... home.

— Edgar Mitchell, Apollo 14, January 1971

Earth the Water Planet

Viewed from space, Earth's oceans and swirling white clouds give it the appearance of a big blue and white marble. In fact, a more appropriate name for our planet might be "Water" because more than two-thirds of its surface is covered by liquid water, a characteristic that makes Earth unique among planets and allows life to flourish here. Water is a key component of the Earth system, flowing between its oceans, rivers, air, rocks, and living things. Water also transports energy from one part of the Earth system to another, and is an important driver of climate change.

The **hydrosphere** is Earth's water in all of its forms: liquid water, vapor, and ice. In this module, you'll learn where Earth's water is stored and how it's constantly cycled from one place to another. You'll also gain an appreciation for how closely the hydrosphere is linked or "coupled" with the atmosphere. This is so much the case that it is hard to separate discussion of these two spheres, and you'll see more examples of the complex interactions between them in the Atmosphere Module of this *Explorer's Guide*.

This module contains several readings, questions, and problems to assist you in your study of the hydrosphere. Your instructor may choose to have you do all or selected parts of this module.

Learning Objectives

After completing the assignments in this module, you should be able to answer the following questions:

1. How do ice cores record ancient climates?
2. Why is the sea salty?
3. How do clouds influence climate?
4. How did Earth's oceans originate?
5. What role(s) do oceans play in the carbon cycle?
6. How do glaciers contribute to both the hydrologic and rock cycles?
7. How does the hydrologic cycle work?

8. What role(s) does water play in Earth's energy budget?
9. How do oceans influence climate?
10. What is El Niño, what are its impacts, and how do scientists study it?

Part I. Getting Started: KWL Chart

Before you begin your study of Earth's hydrosphere, take a few moments to fill out the three tables below. You will come back to them at the end of this module to see what you've accomplished, check if your questions have been answered, and ask any new questions resulting from your investigations.

TABLE 3.1   KWL chart about rivers.

K What I Know	W What I Want to Know	L What I Learned

TABLE 3.2   KWL chart about glaciers.

K What I Know	W What I Want to Know	L What I Learned

TABLE 3.3   KWL chart about oceans.

K What I Know	W What I Want to Know	L What I Learned

Part II. Earth System Snapshots

Icy Archives

Themes: Interactions of Spheres; Humans and the Earth System

How much are humans changing the climate by adding carbon dioxide and other **greenhouse gases** to the atmosphere? To answer this question, we need to know how CO_2 levels have changed through history, both before and after humans began burning huge quantities of fossil fuel. But, direct measurements of carbon dioxide levels in the atmosphere have only been made since 1958. That doesn't even get us back to the Industrial Revolution of the 1850s. But there is a way to sample air that was in the atmosphere hundreds or even thousands of years ago: drill into the thick ice sheets of Antarctica and Greenland.

Why are polar ice sheets a kind of "museum" for old air? Glaciers grow by the accumulation of layers of snow, a relatively fluffy material mixed with air. As more snow falls, it compresses underlying layers, forming ice which traps some bubbles of air. The ice also captures any layers of dust that settled on top of the snow. Each year, more snow falls, leading to the accumulation, through time, of thick sheets of ice. Annual layers of ice can be counted much like tree rings to derive a "calendar" for events recorded in an ice core (Figure 3.1).

Scientists have drilled deep into the ice sheets of Greenland and Antarctica, extracting cores that represent thousands of years of precipitation. In Greenland, cores more than 2 km long record greater than 160,000 years of climate history. Several important types of information are derived from these icy records:

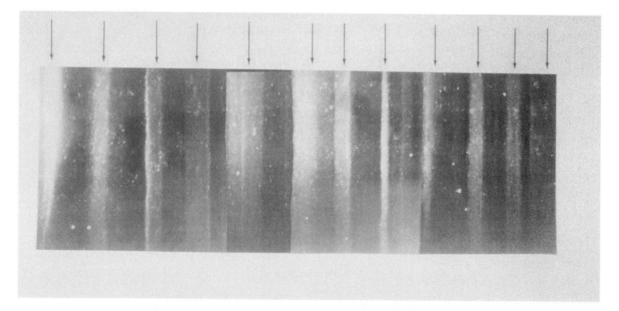

Figure 3.1 Summer and winter layers in a core taken from the Greenland ice sheet. This 19-cm long section shows eleven annual layers, with summer layers (arrowed) sandwiched between darker winter layers. (*Source*: National Oceanic and Atmospheric Administration.)

1. Composition of ancient atmosphere, including concentrations of carbon dioxide and methane, two important **greenhouse gases**.
2. Data about past winds. The size of particles preserved as dust layers within the ice tells of wind strength because stronger winds can carry larger particles.
3. Volcanic eruptions leave two types of signatures in glacial ice: layers of ash and increased acidity of ice due to deposition of volcanic **aerosols**.
4. **Paleotemperature** can be estimated by measuring the ratio of two **isotopes** of oxygen, oxygen-18 and oxygen-16, in the ice. This works because the relative proportions of these two isotopes in precipitation changes as air temperature changes.

Why is the Ocean Salty?

Themes: Interactions of Spheres

Seawater is a complex solution consisting, on the average, of about 965 parts water to 35 parts salts. A look at Chapter 12 in *Earth Science* reveals that Na^+ and Cl^- are the most abundant ions in seawater, which also contains K^+, Mg^{2+}, Ca^{2+}, SO_4^{2-}, HCO_3^-, and minor amounts of many other elements. One source of sea salts is **chemical weathering** of continental rocks such as granite. For example, feldspar, a mineral found in granite, weathers as follows:

$$2NaAlSi_3O_8 + 2(H^+ + HCO_3^-) + H_2O = Al_2Si_2O_5(OH)_2 + 2Na^+ + 2HCO_3^- + 4SiO_2$$

| Sodium feldspar | Carbonic acid | Water | Clay mineral | Sodium ion | Bicarbonate ion | Silica |

Rivers carry the sodium and bicarbonate ions and the silica to the sea.

Although weathering of rocks on the continents supplies most of the materials dissolved in seawater, it is not the only source. At *hydrothermal vents* (Box 13.2 on p. 367 in *Earth Science*) seawater circulates through hot, newly formed ocean crust, adding some elements to the water and subtracting others. A third source of material is underwater volcanic eruptions that transfer material from Earth's mantle to the oceans via **outgassing**. Negatively charged ions in seawater such as Cl^-, SO_4^{2-}, and Br^- are contributed by outgassing and are significantly more abundant in the ocean than in Earth's crust.

So, if rivers, hydrothermal activity, and volcanism are continually adding new salts to the oceans, shouldn't the oceans be getting saltier and saltier? They're not. In fact, the oceans have had about the same average salinity for hundreds of millions of years. In other words, the ocean appears to be in a **steady-state equilibrium** with a balance between the addition of salts and their removal. Several processes remove ions from seawater including extraction of calcium and carbon dioxide by creatures that make their shells of calcite ($CaCO_3$), inorganic precipitation of rocks and minerals such as limestone and rock salt (halite), and chemical reaction of hot oceanic crust with cold ocean water.

Clouding the Issue

Themes: Interactions of Spheres; Energy

It seems so simple. When asked how clouds affect temperature, most people would say that they cool things down by blocking the Sun's rays. You've probably had the experience of enjoying a cool but sunny day until the clouds rolled in and sent you scurrying after a sweater. But, like many things that seem simple, the role of clouds in climate is anything but simple. In fact, because clouds can have both the effects of cooling and warming the atmosphere, they present climate modelers with a real dilemma and this uncertainty about clouds has fueled debates about global warming. How can clouds have different and opposite effects? The answer lies in understanding how clouds interact with energy coming in from the Sun and radiating outward from the Earth.

All objects emit radiant energy, but the hotter the body, the shorter the **wavelength** of its maximum radiation. The Sun, with a surface temperature of nearly 6,000 degrees C radiates its maximum energy in the **visible** range (pp. 418–419 and Figure 15.15 in *Earth Science*). Some of that energy is reflected back into space and some is absorbed by oceans and land. Most of the absorbed energy is re-radiated to space. Because Earth is much cooler than the Sun, the radiation it emits is of longer wavelength (**infrared**; see Figure 15.15 in *Earth Science*). Earth's atmosphere lets shorter-wavelength solar radiation in, but certain gases in the atmosphere (**greenhouse gases**) absorb the longer-wavelength radiation emitted by the Earth and thus prevent it from escaping to space. This phenomenon is known as the **greenhouse effect** (see p. 421 in *Earth Science*), but the analogy is imperfect because a real greenhouse is warmed by a different mechanism. The greenhouse glass lets solar radiation in, but prevents loss of heat by not allowing the warmed air inside to mix with cooler air outside of the greenhouse.

Because clouds have a high **albedo** or reflectivity, they reflect incoming solar radiation and cause planetary cooling. But, because water vapor is a very effective greenhouse gas, clouds trap out-going radiation emitted from the Earth. This leads to warming. The effect of a particular cloud depends on its height and thickness (Figure 3.2). Because they are transparent to incoming shortwave solar radiation but can readily absorb outgoing longwave radiation, the overall effect of high, thin **cirrus** clouds is to enhance greenhouse warming. In contrast, low *stratocumulus* clouds (p. 451 in *Earth Science*) are thick and not as transparent as cirrus clouds. Although they do trap infrared radiation emitted from the Earth, these clouds have a net cooling effect because they reflect a lot of incoming shortwave radiation.

Because clouds are highly variable in their distribution and properties and have both cooling and warming effects, they have been called a "wild card" in climate models. The NASA satellite TERRA, launched in 1999, is gathering data about clouds to help scientists understand how they contribute to climate change.

Volcanoes and Origin of Earth's Water

Themes: Interactions of Spheres

As discussed in the Geosphere Module of this workbook and on pp. 230–231 of *Earth Science*, the most abundant gas emitted by volcanoes is water vapor. In fact, most scientists believe that, with the possible exception of some delivery of water from comets, it was volcanism that formed Earth's oceans (and atmosphere) billions of years ago.

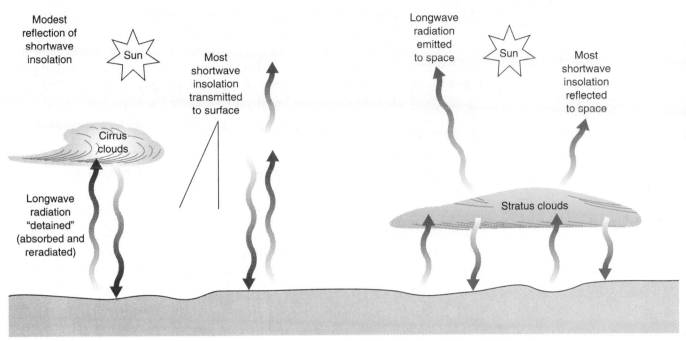

Figure 3.2 Energy effects of high and low clouds. a. High, thin cirrus clouds transmit most of the incoming shortwave radiation from the Sun, but absorb and delay loss of outgoing longwave (infrared) radiation. These clouds have a net warming effect. b. Low, thick stratocumulus clouds reflect most of the incoming shortwave radiation and radiate longwave radiation to space. These clouds have a net cooling effect.

Earth and the other planets are believed to have formed at about the same time from a huge cloud composed mostly of hydrogen and helium; this is known as the **nebular hypothesis** (see pp. 2–5 in *Earth Science*). About 5 billion years ago, the rotating cloud or **nebula** began to contract, sweeping most of the mass towards the center and producing the Sun. Earth and other planets formed by accretion of material remaining in the flattened, rotating disk. Shortly after Earth's formation, heat from radioactive decay and collision of particles caused melting in the planet's interior. This allowed *differentiation* or density sorting into layers, with settling of the dense iron and nickel core towards Earth's center and migration of less dense material towards the surface. During differentiation, water vapor that was originally contained in the mantle was released to the atmosphere by volcanic eruptions. This process, known as **outgassing**, continues today. With cooling, the water vapor condensed and began accumulating as liquid water.

Oceans and the Carbon Cycle

Themes: Interactions of Spheres; Scale; Cycles; Humans and the Earth System

Oceans interact with the atmosphere in both physical and chemical ways. As you will see in an exercise below, oceans influence climate through the physical processes of heat absorption and transport. Oceans interact chemically with the atmosphere by absorbing and storing large quantities of the **greenhouse gas** carbon dioxide (CO_2); this is an important process in the short-term carbon cycle (pp. 2-18–2-19 and Figure 2.9 in the Geosphere Module of this workbook). Scientists estimate that oceans currently absorb between 30% and 50% of the CO_2 produced by burning fossil fuels and believe that without the oceans, atmospheric CO_2 levels would be much higher (~500–600 **parts per million** or ppm) than today's level of 360 ppm.

Carbon dioxide dissolves readily in seawater, where it is taken up by **phytoplankton** (microscopic plants) that consume it during **photosynthesis**. Most of the gas escapes back to the atmosphere within a year or so, but some is transported to the deep ocean when phytoplankton die and settle to the ocean floor. The transfer of organic carbon to the deep ocean is called the "**biological pump**." The CO_2 is released into the water as the organic material decays and most is absorbed by reacting with water to form carbonate (CO_3^{2-}) and bicarbonate (HCO_3^-) ions, which combine with Ca^{2+} to form calcium carbonate (limestone). Carbonate sediments constitute a vast long-term "storehouse" for carbon dioxide.

One consequence of global warming due to increased levels of human-produced CO_2 in the atmosphere would be warmer oceans. Warm water absorbs less CO_2 than cold water, so warming oceans would lead to more CO_2 left in the atmosphere, which would lead to more warming as part of a **positive feedback** loop.

Part III. Questions

A. Short Answer
1. How does plate tectonics influence the type of shoreline?

2. Describe the interactions between two or more Earth spheres that are responsible for:
   a. Surface ocean currents (pp. 376–380 in *Earth Science*)

   b. **Thermohaline circulation** (p. 381 in *Earth Science*)

   c. Formation of **stalactites**

   d. **Geysers**

3. What roles do glaciers play
   a. in the rock cycle?

   b. in the hydrologic cycle?

4. How does plate tectonics (geosphere) influence formation of glaciers? (Hint: See pp. 146–147 in *Earth Science*).

5. Are glaciers **open systems** or **closed systems**? Explain.

6. What is the relationship between glaciers and sea level?

7. How do you think increased rates of seafloor spreading might affect sea level?

8. Explain the part(s) played by water in:
   a. **Mechanical weathering**:

   b. **Chemical weathering**:

B. Longer Answer

More about systems

The exercises below will introduce you to some additional aspects of systems. As you will recall from Module 1, a **system** is a group of inter-acting parts that forms a complex whole. An **open system** is character-ized by *inputs* of energy and matter from the surroundings and *outputs* of energy and matter leaving the system. When inputs are equal to out-puts, the system is balanced and is said to be in **steady-state equilib-rium**. In this type of equilibrium, the average condition of the system stays the same. For example, as you read in the "Why is the Ocean Salty?" Snapshot above, the ocean is in a steady state with respect to its average salinity, which has stayed more or less the same for hundreds of millions or possibly billions of years.

If a system at equilibrium experiences a change in input or output, it must respond in a way that will restore equilibrium. This is where **feedback** comes in (see Module 1). Feedback allows the system to reg-ulate itself.

To aid their study of the complex Earth system, scientists construct *computer models* that attempt to understand, in a quantitative fashion, cycles that transfer a specific material such as water or carbon among interconnected **reservoirs**. The components of such models include:

1. Identification of the **reservoirs** or storage containers for the sub-stance in question;
2. *Flux*, or the amount of the substance entering or leaving the reser-voir in a given time period;
3. **Residence time**, or the average time that a substance remains in a reservoir; and
4. *Budget*, or the balance sheet of all inputs and outputs to a reser-voir or series of interconnected reservoirs.

TABLE 3.4 Reservoirs of the hydrosphere
and their volumes.

Reservoir	Volume (10^6 km^3)
Oceans	1400
Ice caps & glaciers	43.4
Groundwater	15.3
Lakes	0.125
Soils	0.065
Atmosphere	0.0155
Rivers	0.0017
Biosphere	0.002
Total	**1459**

Models of the Earth system are simplifications of a complex reality, but they are nonetheless powerful tools that help scientists predict the system's response to change. For example, models of the current climate system are attempting to answer such questions as "What will happen if humans continue to put more and more carbon dioxide into the atmosphere?"

1. **Earth's Water Budget**

You're familiar with the concept of balancing (or trying to balance) your budget. If your income exceeds your expenses, you are able to save money. If you spend more than you earn, you go into debt. If income equals expenses, the budget is in balance. You might have different storage places for your money: a savings account, a checking account, and cash in your sock drawer. In considering your total assets, you add up the amounts in these different places. You might decide to move your funds from one place to another, but the amount will remain the same as long as you don't add to or subtract from the total sum.

We can take a similar budget analysis approach to modeling cycles of matter in the Earth system by looking at storage places, total assets, and flow from one place to another. Earth is an essentially **closed system** with respect to water; that is apart from minor loss to and gain from the **cosmosphere**, the total amount of water on Earth is constant: about 1.46 billion cubic kilometers (km^3). Table 3.4 shows the distribution of water among different **reservoirs** or storage places.

a. What is the largest reservoir for water on Earth?

b. What is the largest fresh-water reservoir?

c. What is the largest fresh-water reservoir that is readily available for use by people?

d. Water on Earth doesn't stay put in any one reservoir, but is constantly moving from one place to another as part of the *hydrologic cycle* (pp. 98–99 in *Earth Science*). Table 3.5 summarizes the flow or flux of water in the Earth system. Please note that the numbers given in Table 3.5 differ somewhat from those shown on Fig. 4.2 in *Earth Science*. For this exercise refer to the fluxes given in Table 3.5.

1. What process or processes *remove* water from the oceans?

TABLE 3.5 Fluxes in the hydrosphere.

Process	Amount of water moved (km³/year)
Evaporation from oceans	434,000
Precipitation into oceans	398,000
Runoff into oceans	36,000
Evaporation from land	71,000
Precipitation onto land	107,000

2. What is the total amount of water withdrawn from the oceans each year?

3. What process or processes *add* water to the oceans?

4. What is the total amount of water added to the oceans each year?

5. How do the inputs and outputs compare?
   What does this imply about the volume of the ocean through time?

6. What change(s) in other parts of the Earth system could cause the volume of the oceans to change?

e. How long does a drop of water stay in a given reservoir?
   To get a sense of how rapidly water moves through different reservoirs, we can calculate a value known as **residence time**:

$$\text{Residence time} = \frac{\text{Total volume in the reservoir}}{\text{Rate of influx or outflow.}}$$

For example, the residence time of water in the atmosphere

$$= \frac{\text{volume of water in the atmosphere}}{\text{inflow (or outflow) of water into the atmosphere}}$$

$$= \frac{0.0155 \times 10^6 \, \text{km}^3}{505,000 \, \text{km}^3/\text{year (total precipitation or evaporation)}}$$

$$= 0.03 \text{ year} \times 365 \text{ days/year} = 11 \text{ days}$$

1. Use the data in Tables 3.4 and 3.5 to calculate a residence time for water in the oceans. *Show your work.

   Fill in your answer in the appropriate place in Table 3.6.

2. How does the residence time of water in the oceans compare to that of water in the atmosphere?

3. According to Table 3.6, how does the residence time of water in rivers compare to that of water stored as groundwater?

4. Which do you think is easier to clean up, stream pollution or groundwater pollution? Why?

2. **Water and Energy**

   In addition to constantly moving water, dissolved minerals, and nutrients from one sphere to another, the hydrologic cycle moves energy from one part of the Earth system to another. Because of the high **specific heat** of water, oceans store vast amounts of energy. Oceans, like the atmosphere, play a major role in maintaining Earth's energy balance by transporting heat from the equator to the poles (see pp. 379–380 in *Earth Science*).

   In the previous exercise, you saw that Earth's water budget is balanced. The same is true of energy: on the average, solar radiation entering the Earth system is balanced by the amount of radiation

TABLE 3.6 Residence times in various reservoirs of the hydrosphere.

Reservoir	Residence time
Oceans	
Ice caps & glaciers	10–10,000 years
Groundwater	2 weeks –10,000 years
Lakes	~ 10 years
Soils	2 weeks –1 year
Atmosphere	~ 10 days
Rivers	~ 2 weeks
Biosphere	~ 1 week

flowing from Earth to space. If this were not true, Earth would be growing hotter or cooler through time. Of course, there are a number of factors on Earth and in the **cosmosphere** (such as increased CO_2 in the atmosphere and changes in Earth's orbit) that can alter Earth's energy budget, leading to climate change.

a. *The hydrosphere and reflection*

The energy that drives circulation of Earth's water and air comes from the Sun. Figure 3.3 shows Earth's energy budget. Note that incoming solar energy is equal to the total amount of energy returned to space by reflection and radiation. As shown in the diagram, 30% of sunlight is reflected back to space.

1. Clouds account for what percentage of reflected solar radiation?

    *Show your work.

2. How would Earth's average temperature change if there were no clouds?

EARTH'S ENERGY BUDGET

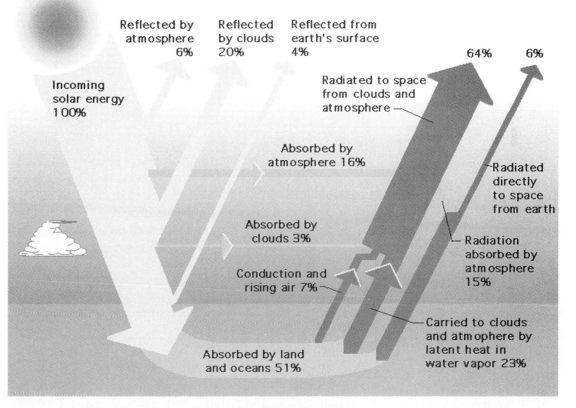

Figure 3.3 Earth's energy budget. (*Source*: NASA Headquarters.)

3. **Albedo** refers to the reflectivity of a surface, usually expressed as a percent of incoming radiation that is reflected. Some natural and artificial materials are better reflectors than others. Look at the albedo of various surfaces listed in Table 15.2 on p. 421 of *Earth Science.*
   a. Which of the surfaces listed has the highest albedo?
   b. The lowest?
4. Referring again to Table 15.2, if global warming were to melt glacial ice, how would the Earth's radiation budget be affected?

   a. Would this change in the radiation budget lead to more or less warming?
      Why?

   b. Is this an example of **positive feedback** or of **negative feedback**?
      Explain.

b. *The hydrosphere and absorption* (see pp. 423–425 *in Earth Science*)
   1. Refer to Figure 3.3. What percentage of incoming solar radiation is absorbed by land and oceans?
   2. Based on what you know about the **specific heats** of water and land, which do you think absorbs more radiation?
   3. How do you think Earth's temperature would change if 71% of its surface were covered by land instead of by water?

3. **Oceans and Climate**
   The climate of a given city is determined by several variables including its: 1. latitude; 2. proximity to an ocean; and 3. location with respect to warm or cold ocean currents (see Fig. 14.2 on p. 377 in *Earth Science*). You'll need an atlas for this exercise.
   a. Use the data in Table 3.7 to make plots of monthly temperatures for cities A, B, C, and D on Figure 3.4. The vertical axis is temperature;

TABLE 3.7 Monthly temperature data for four cities.

						Monthly temperature (°C)							
City	Latitude	Jan	Feb	Mar	Apr	May	Jun	Jul	Aug	Sep	Oct	Nov	Dec
A	59°N	−27.7	−26.4	−20.5	−10.1	−1.5	5.9	12.1	11.3	5	−1.8	−12.9	−22.7
B	60°N	−6.1	−6.5	−3.5	2	8.5	14	16.8	15.5	5	5.5	0.4	−3.5
C	48°N	4.1	5.9		9.5	12.9	15.7	18.1	18.1	15.6	11.3	7.1	4.7
D	47°N	−12.8	−10.7	−3.3	6.1	12.8	18	21.5	20.3	14.2	7.3	−2.1	−9.1

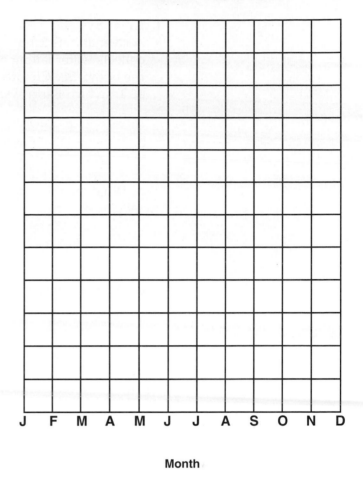

J F M A M J J A S O N D

Figure 3.4 Graph of monthly temperature data for
cities A, B, C, and D.

Month

you will need to label it with the appropriate increments. Use a
different color or line thickness for each city.

b. Cities A and B are at about the same latitude, so variation in solar
   radiation with latitude doesn't explain the difference between
   them. One of these cities Is Helsinki, Finland, and the other is
   Churchill, Canada. Based on your graph which city, A or B, do
   you think is Helsinki? Explain your reasoning.

c. Cities C and D are also at about the same latitude, eliminating difference in latitude as an explanation for their different climates.
   One of these cities is Seattle, Washington, and the other is Bismarck, North Dakota. Based on your graph which city, C or D, do
   you think is Bismarck? Explain your reasoning.

4. **River Systems**

A river and its tributaries may be thought of as a natural system that collects water falling as precipitation on land and delivers it back to the ocean. See Chapter 4 in *Earth Science*.

a. To which Earth system cycle(s) do rivers contribute?

b. What are the inputs to a river system?

c. What are the outputs of a river system?

d. What energy source(s) drive a river system?

e. What work is done by a river system?

Part IV: Concept Map About Rivers

Fill in the question marks in the concept map shown in Figure 3.5. Add three new concepts and show how they link with other concepts.

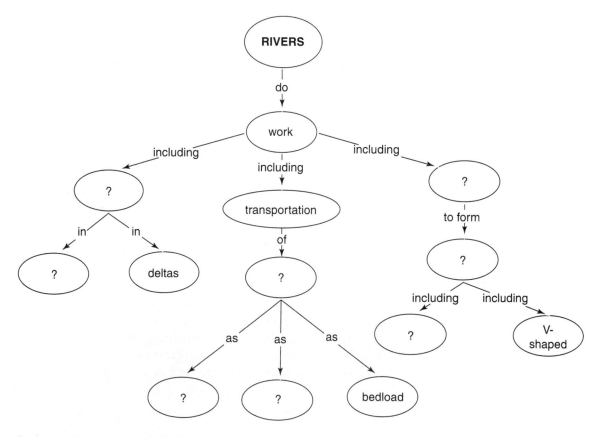

Figure 3.5 Concept map about rivers.

Part V. WebQuest - El Niño News

Introduction

How many times has **El Niño** been blamed for bad weather? But, what exactly is El Niño and why is it often held responsible for our weather woes? In this WebQuest, you'll learn how Earth's hydrosphere and atmosphere interact to generate an El Niño, discover how scientists study this phenomenon, and explore its regional and global consequences.

The Task

You've been hired by the National Oceanic and Atmospheric Administration (NOAA) and the National Aeronautics and Space Administration (NASA), both of whom are worried that they will be the next targets of budget cuts, to help convince the public that it's important to support ongoing studies of El Niño. NOAA and NASA want you to write a brief article (your instructor will specify the length) for your campus newspaper that will address the following:

- What is El Niño?
- How do scientists study it?
- What are the social and economic impacts of El Niño?
- What are the benefits of predicting it?

The Process

First, you will need to do some research. Use information obtained from the Web sites below to clarify your own understanding of the four areas you are to address in your article. Take notes. Then, write your article *using your own words* to explain the points that NOAA and NASA have asked you to include. Make sure that your article is appropriate for your target audience: the educated layperson of the campus community who doesn't necessarily have an Earth science background.

Resources

NOAA's El Niño Theme Page
http://www.pmel.noaa.gov/tao/elnino/nino-home.html
NOVA On-line: Tracking El Niño
http://www.pbs.org/wgbh/nova/elnino/
Ocean Surface Topography from Space
http://topex-www.jpl.nasa.gov/science/el-nino.html
University of Illinois Weather World 2010 Project
http://ww2010.atmos.uiuc.edu/(Gh)/guides/mtr/eln/home.rxml

Learning Advice

- Did you use information from multiple sources?
- Did you crosscheck the information you found?
- Did you compare the information from different sources?

Conclusion

What do we know about El Niño and why is it important to continue our investigations of this phenomenon?

Part VI. Summing Up the Sphere

Using information from the "Earth System Snapshots," "Earth as a System" boxes in your text, and your own ideas, describe interactions of the hydrosphere with Earth's other components. Give at least two examples of each of the following interactions:

Hydrosphere-Atmosphere:

Hydrosphere-Biosphere:

Hydrosphere-Geosphere:

You may also think of examples that involve more than two spheres. Your examples:

Part VII. KWL Revisited

Go back to the KWL charts you filled in at the beginning of this module. Did you have any initial misconceptions about the topics you've explored? Go back to correct them. Do you have any unanswered questions?

Module 4

Atmosphere

The Atmosphere

With every breath you take, you are a living example of the interconnectedness of the **biosphere** and the **atmosphere**. The atmosphere is a thin layer of gases that envelops our planet. If Earth were a peach, the atmosphere would be only about as thick as the fuzz, but don't let its thinness mislead you about its importance. Without the heat-trapping and radiation-shielding effects of the atmosphere, life as we know it could not exist on our planet.

Learning Objectives

After completing the assignments in this module, you should be able to answer the following questions:

1. How do tree rings record climatic conditions of the past?
2. How do volcanoes affect climate?
3. How does life influence climate?
4. What is the **Gaia hypothesis**?
5. What natural processes convert atmospheric nitrogen to compounds that living things can use?
6. How do humans interfere with the **nitrogen cycle**? What are some environmental consequences of this interference?
7. How can plate tectonics cause climate change?
8. What variations in Earth's orbit cause climate change on a scale of thousands of years?
9. Give two examples of **feedback** between vegetation and climate.
10. Describe the composition of Earth's atmosphere.
11. What was the origin of the atmosphere, and how has it changed through time?
12. How does Earth's atmosphere differ from those of Venus and Mars? What accounts for the differences?
13. What is global warming, and why is this such a controversial issue?
14. Explain how global warming can affect sea level.
15. Explain how the greenhouse effect works. What gases are involved?
16. What human activities are contributing to rising concentrations of CO_2 and other **greenhouse gases**?

TABLE 4.1   KWL chart about Earth's atmosphere and climate.

K What I <u>K</u>now	W What I <u>W</u>ant to Know	L What I <u>L</u>earned

Part I. Getting Started: KWL Chart

This module will focus on the origin and evolving composition of the atmosphere and on natural and human-caused controls on climate change. Before you begin your study of Earth's atmosphere and climate, take a few moments to fill out Table 4.1. You will come back to it at the end of this module to see what you've accomplished, check if your questions have been answered, and ask any new questions arising from your investigations.

Part II. Earth System Snapshots

The Tales that Tree Rings Tell

Themes: Interactions of Spheres

Your everyday experience tells you that plants respond to changes in their environment. They droop if you don't give them enough water and wilt if they get too much sunlight. Because moisture and temperature influence tree growth, scientists can use the rings that record annual growth to study past climates. Tree rings are just one example of how scientists use **proxy data** (data from objects that are sensitive to climatic variables such as temperature and precipitation) to study **paleoclimate** (climate of the past). Other examples of proxy data are ice cores, marine sediments, and fossil pollen.

How do tree ring studies work? Each year, trees add to their width, with the new growth referred to as a *tree ring* (Figure 4.1). These growth layers are revealed in the cross section of a tree trunk when the tree is cut down or obtained by using a borer that's screwed into the tree and pulled out, bringing with it a straw-sized core sample of wood. The rings can be counted to determine the age of the tree, which may be hundreds or even thousands of years. Most trees grow more during wet, cool years than during dry or very hot years. This is reflected in wider rings than are formed during times of drought.

An interesting story of how tree rings helped to solve a sixteenth-century historical mystery comes from Roanoke Island, North Carolina, where, in 1590, more than a hundred English colonists disappeared without a trace. Tree-ring data indicate that the disappearance of the "Lost Colony" coincided with the most extreme drought in 800 years. A few years later, colonists in Jamestown, Virginia, suffered through a famine known as the "Starving Time." Evidence from tree rings shows that the settlers had the extraordinarily bad luck of arriving in the midst of the driest seven-year period in 770 years. Tree-ring data also suggest that drought was responsible for the abandonment, some 1,000 years ago, of Anasazi cliff dwellings in the American Southwest. You can read more about tree rings in Box 10.3 on pp. 302–303 in *Earth Science*.

Volcanoes and Climate (See also Box 8.2 on p. 253 in *Earth Science*.)

Themes: Interactions of Spheres

You probably know Ben Franklin for his role as a Founding Father and his legendary kite and key experiment. You might not know that he was also the first person to suggest a connection between volcanoes and climate change. In 1783–1784 when Franklin was serving as U.S. Ambassador to France, Europe experienced a pervasive "dry fog" and an unusually severe winter. He suggested that the fog and cold weather were due to the eruption, in 1783, of Laki Volcano in Iceland.

Figure 4.1 Tree rings
(*Source*: Corbis/Stock Market.)

In the more than 200 years since then, global cooling has been observed to follow a number of volcanic eruptions. The cataclysmic eruption, in 1815, of Tambora Volcano in Indonesia lowered global temperatures by as much as 3 °C. The following year was known as the "year without a summer," and was marked by June snows in New England. Other examples of gigantic blasts that have been implicated in global cooling are Kratakota (Indonesia, 1883), Santa María (Guatemala, 1902), and Agung (Indonesia, 1963).

Volcanoes can cause cooling by ejecting tons of dust and sulfur dioxide gas into the atmosphere. Solar radiation is blocked by the dust and **aerosols** (tiny particles that form when SO_2 reacts with water to form droplets of sulfuric acid). In a few days, rain and snow usually wash out particles that reach the lower part of Earth's atmosphere (the **troposphere**). In contrast, eruptions that are explosive enough can launch material high into the **stratosphere**, where it may circle the globe for several years.

From 1980 to 1991, a trio of major eruptions provided scientists with an opportunity to learn more about the connection between volcanoes and climate. Satellite monitoring made it possible to track clouds of material ejected from these volcanoes, and to measure temperature changes following their eruptions.

The 1980 eruption of Mt. St. Helens in Washington State was one of the largest of the 20th century, but it had virtually no impact on climate. This is because Mt. St. Helens emitted little sulfur and its sideways blast prevented material from reaching the stratosphere.

Two years later, in 1982, Mexico's El Chichón erupted. Although Mt. St. Helens pumped much more ash into the atmosphere, El Chichón emitted forty times as much sulfur-rich gas. El Chichón lowered temperatures by about 0.3 °C, but the overall cooling may have been greater if its eruption had not coincided with a warmth-producing El Niño event.

In 1991, Mount Pinatubo blasted 20 million tons of sulfur compounds into Earth's stratosphere. Global temperature dropped by approximately 0.5 °C for about two years after the eruption.

Does Life Control Climate? An Introduction to the Gaia Hypothesis

Themes: Interactions of Spheres

This workbook includes many examples of how the biosphere interacts with the non-living parts of the Earth system. We tend to think in terms of life being at the mercy of the physical environment. But a hypothesis put forth by biologists James Lovelock and Lynn Margulis states that life actually manipulates the physical and chemical environment on a global scale for its own benefit. This idea, which is called the **Gaia hypothesis** after the Greek goddess of mother Earth, has provoked much discussion and controversy. Most scientists have rejected the so-called "strong" Gaia hypothesis with its implication that the Earth is an actual organism that purposefully regulates the environment, but many other aspects of the idea are being investigated.

Observations that led to the Gaia hypothesis are: 1. Earth's atmosphere is strikingly different in composition from the atmospheres of its lifeless neighbors Venus and Mars; and 2. Earth's temperature has remained within the narrow range required for life even though the Sun has grown approximately 25% brighter since the formation of the solar system.

What role does life play in atmospheric chemistry and in keeping the planet from overheating?

Earth's early atmosphere is thought to have formed as the result of volcanic **outgassing** (see pp. 310–311 in *Earth Science*) which brought volatiles to the surface from the planet's interior. In contrast to today's atmosphere, which is 21% oxygen, the ancient atmosphere would have had almost no oxygen. The rise of atmospheric oxygen on Earth and its near absence on Mars and Venus can be explained by the presence of life on Earth and its absence on the other worlds. Through **photosynthesis**, green plants use solar energy to manufacture food from carbon dioxide and water:

$$6CO_2 \quad + \quad 6H_2O \quad = \quad C_6H_{12}O_6 \quad + \quad 6O_2$$

Carbon + Water $\qquad$ Glucose + Oxygen
dioxide

Oxygen is a byproduct of photosynthesis. Primitive marine algae capable of photosynthesis appeared on Earth by about 3.5 billion years ago so life's influence on atmosphere began early in the planet's history, although it took many millions of years for oxygen to build up to its current 21%.

One way that organisms can influence Earth's surface temperature is by affecting levels of CO_2 in the atmosphere. Carbon dioxide is an important **greenhouse gas**, and increases in atmospheric CO_2 cause increases in temperature. Carbon dioxide is added to and removed from the atmosphere as a result of various biologic processes (see Figure 2.9 on p. 2-18 of this workbook). Extraction of CO_2 occurs as a result of photosynthesis and production of calcium carbonate ($CaCO_3$) shells. Addition of CO_2 occurs through **respiration** (the reverse of photosynthesis) and decay of organic matter. The moderation of climate by organisms involves a series of **feedback** loops. For example, rising temperature leads to greater **phytoplankton** (p. 362 in *Earth Science*) production, which leads to more removal of CO_2 from the atmosphere, which leads to cooling. Another mechanism for temperature control is the production by marine phytoplankton of a compound called *dimethyl sulfide (DMS)*. In the atmosphere, DMS oxidizes to form **aerosols** of sulfate salts, which serve as nuclei for cloud formation. By reflecting solar radiation, clouds can cool the Earth. As noted above, an increase in temperature stimulates growth of phytoplankton, which through generation of DMS promotes cooling in a **negative feedback** loop. This regulation of climate might remind you of the thermostat you set to keep your home at a comfortable temperature.

Critics of the Gaia hypothesis maintain that although Earth's biosphere interacts with the inanimate parts of the Earth system, climate regulation is primarily the result of physical and chemical processes such as silicate weathering, plate tectonics and, volcanism (see discussion of the **long-term carbon cycle** in Module 2). Although there is much disagreement about the validity of the Gaia hypothesis, it has had the positive effect of stimulating discussion of Earth as a single integrated system.

Nitrogen, Nitrogen Everywhere and Not an Atom to "Eat" ...

Themes: Interactions of Spheres; Cycles; Humans and the Earth System

It may not be as obvious as the need for water, but all living things require nitrogen (N_2) to make such key organic molecules as amino acids, proteins, and genetic material. Because Earth's atmosphere is composed of

78% N_2 gas, it is an abundant source of nitrogen, but most organisms can't make direct use of it. The bonds in N_2 are very strong, making it highly unreactive. Before living things can use nitrogen, it must first be "fixed," meaning that it must be combined with oxygen, carbon, or hydrogen to make reactive compounds that organisms can utilize. Recall from earlier discussions that the Earth system is characterized by a number of *cycles*, including the hydrologic, rock, and carbon cycles. The flow of nitrogen from the **atmosphere**, to the soil (**geosphere**), to the **biosphere** and back to the atmosphere is another example of the constant cycling of elements among different **reservoirs** in the Earth system.

In nature, certain forms of bacteria fix nitrogen. Symbiotic nitrogen-fixing bacteria live in nodules on the roots of peas, beans, peanuts, alfalfa, and other legumes. Although it produces less fixed nitrogen than do biological processes, lightning is another player in the **nitrogen cycle** (and you thought the only thing that lightning is good for is a great light show). The high temperature and pressure in the air through which a flash of lightning travels causes nitrogen (N_2) and oxygen (O_2) molecules to break up and recombine as nitrogen oxides. Nitrogen oxides combine with water in the atmosphere to form *nitric acid* (HNO_3), which breaks down to produce hydrogen ions and nitrate (NO_3^-), a usable plant nutrient that is washed into the soil when it rains.

Humans are interfering with the natural nitrogen cycle through combustion of fossil fuels in power plants and car engines, industrial production of nitrogen-based fertilizers, and planting of legumes which host nitrogen-fixing bacteria. It is estimated that humans are now responsible for more than 50% of the planet's fixed nitrogen. **Anthropogenic** (related to human activities) nitrogen fixation leads to several environmental problems. High-temperature combustion of fossil fuels introduces nitrogen oxides into the atmosphere, where they enhance the greenhouse effect and contribute to the formation of acid rain and smog. Nitrogen-based fertilizers that wash from fields to ponds, lakes, and coastal waters may cause algal blooms. Subsequent decomposition of the algae extracts oxygen from the water and reduces or eliminates populations of fish and other aquatic organisms. High levels of nitrates in drinking water can cause life-threatening illness in babies ("blue baby" disease) and have been linked to some cancers.

Part III. Questions

A. Short Answer

1. First review the discussion of causes of glaciation found on pp. 143–148 in *Earth Science*.
   a. How can plate tectonics cause climate change?

b. On what time scale does climate change brought about by moving plates operate?

c. What three factors did Milankovitch identify as contributing to shorter-term climate change? Briefly describe each factor and, referring to Figure 5.23 on p. 149 in *Earth Science*, indicate the length of cycle involved in each.

2. Vegetation-climate feedback

a. Where on Earth are **tundra climates** found? (See p. 536 in *Earth Science*.)

b. Some of the plants that grow in treeless areas with tundra climates are lichen and dwarfed shrubs.

1. What change in an area's climate would produce a transition from tundra vegetation to forest?

2. Given that tundra plants have a higher **albedo** (see p. 420 in *Earth Science*) than forest vegetation, how would the replacement of tundra by forest further affect climate?
Is this an example of **positive** or of **negative feedback**?
Explain.

c. When climate becomes wetter, forests replace grasslands. Trees put more water vapor into the atmosphere by **transpiration** than does grass. Increased transpiration leads to more precipitation. Is this an example of **positive** or of **negative feedback**?
Explain.

3. What is the source of energy that drives weather and climate?

B. Longer Answer

1. **Climate change and sea level rise**

   a. How much has sea level risen in the past century? (See p. 544 in *Earth Science*.)

   b. How much do some investigators predict it will rise in the next century?

   c. Referring to the map shown in Figure 5.3 on p. 133 in *Earth Science*, name four cities that would be underwater if present ice sheets in Greenland and Antarctica were to melt.

   d. Calculate rise in sea level

   Melting and freezing of polar ice are good indicators of global climate change, so scientists from the National Aeronautics and Space Administration (NASA), National Oceanographic and Atmospheric Administration (NOAA), and the Department of Defense are using satellites to measure changes in our planet's ice sheets. Because it is at a lower latitude than the Antarctic ice sheet, the Greenland ice sheet is more vulnerable to melting as a result of global warming. How much would sea level rise if the Greenland ice sheet melted? Table 4.2 shows the volume of ice in present-day reservoirs.

   To calculate the rise in sea level due to melting of ice, you need to know two things:

   - The area of the ocean ($361,000,000 \text{ km}^2$), and
   - How density changes in the conversion of ice to water. Ice is less dense than water (think of ice cubes floating in your soda). The conversion factor is 0.9.

   For example, to figure out how much sea level would rise as a result of melting of the East Antarctic ice sheet, divide the volume of ice by the area of the ocean:

   $$\frac{26,039,200 \text{ km}^3}{361,000,000 \text{ km}^2} = 0.072 \text{ km}$$

   Now multiply this number by 0.9:

   $$0.072 \text{ km} \times 0.9 = 0.0648 \text{ km} \times \frac{1,000 \text{ m}}{1 \text{ km}} = 64.8 \text{ m}$$

TABLE 4.2   Volumes of Earth's ice sheets.

Location	Volume (km³)
East Antarctic Ice Sheet	26,039,200
West Antarctic Ice Sheet	3,262,000
Antarctic Peninsula	227,100
Greenland	2,620,000
All other ice caps, ice fields, and valley glaciers	180,000
Total	**32,328,000**

Calculate how much sea level would increase if the Greenland ice sheet melted. *Show your work.

e. Is melting of ice the only way that global warming would contribute to changing sea level?
Explain. (Hint: See pp. 543–544 in *Earth Science*.)

Before answering parts f-h, refer to pp. 397–398 in *Earth Science*.

f. On Figure 4.2, indicate which coasts are tectonically quiet and which are active.
g. Write a brief description of the following types of coasts:
Emergent:

Submergent:

h. Based on your answers to questions f and g, which coasts of the United States are most vulnerable to the effects of rising sea level? Explain your answer.

i. Relative changes in sea level may reflect either changes in the level of the ocean or vertical movement of the land surface.

Figure 4.2 Map of the United States.

1. How could plate tectonics cause relative sea level to fall?

2. How do you think increasing rates of seafloor spreading would influence sea level? Explain.

2. **Humans and climate**
   Before answering the following questions, review p. 421 in *Earth Science*.
   a. Write a paragraph using your own words to explain how the **greenhouse effect** works.

   b. Look at Figure 2.9 on p. 2-18 of this workbook. What are two ways that humans add CO_2 to the atmosphere?

   c. Figure 4.3 shows atmospheric levels of CO_2 as measured at the Mauna Loa Observatory in Hawaii. This is the longest record we have of such measurements. Note the zigzag pattern; this reflects seasonal variation in plant growth and photosynthesis.
      The units of measurement are ppmv, which stands for **parts per million** by volume. To convert ppm to the more familiar percent, move the decimal place four places to the left. For example, 300 ppm is the same as 0.03%.
      1. What is the overall trend in CO_2 concentration for the period of time shown?

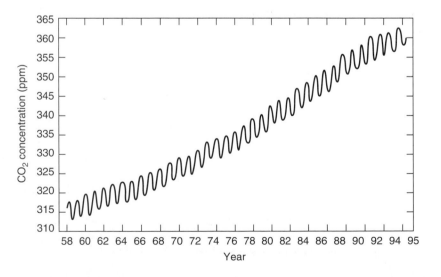

Figure 4.3 Atmospheric CO_2 concentrations measured at the top of Mauna Loa in Hawaii.

2. How many years of direct measurements do we have? (Note: Measurement has continued since 1995.)

d. Clearly, we have direct information about CO_2 for only a very, very tiny fraction of Earth history (recall that Earth is 4.6 billion years old). To gain a perspective on natural climate change and whether human activities are causing global warming, we need to look at fluctuations in CO_2 and temperature over much longer periods of time. By looking at past changes, we may be better able to understand what might happen in the future. How can we determine CO_2 levels for time periods going back centuries or even thousands of years? (Recall the Earth System Snapshot "Icy Archives" in Module 3). Figure 4.4 shows atmospheric CO_2 levels and temperature for the past 150,000 years based on information from the Vostok (Antarctica) ice sheet.

1. Describe how concentration of CO_2 has changed during the interval from 150,000 years ago to the present.

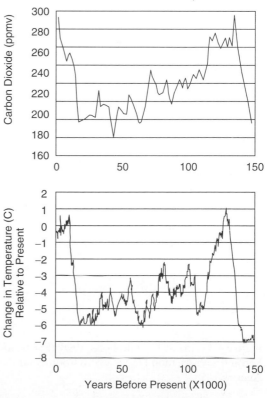

Figure 4.4 Variations in CO_2 concentration (upper curve) and estimated change in temperature relative to the present for the past 150,000 years as determined from air bubbles in an ice core from Vostok, Antarctica. (*Source*: National Oceanic and Atmospheric Administration.)

2. Were there any times in the past when CO_2 was as high as its current level (360 **parts per million**)?

3. What natural processes might account for increased concentrations of atmospheric CO_2? (See pp. 2-18–2-20 in this workbook.)

4. Does there seem to be any relationship between temperature and CO_2? Explain.

e. The United States represents less than 5% of the world's population. According to Figure 4.5, the United States was responsible for what percentage of the world's CO_2 emissions in 1998?

f. Use the Environmental Protection Agency's on-line calculator at **http://www.epa.gov/globalwarming/tools/ghg_calc.html** to estimate your household's annual emissions of carbon dioxide.

My estimate: _____ pounds of carbon dioxide/year

1. How does your calculated emission total compare with the EPA's estimated average of about 60,000 pounds per year for a household of two?

2. List three actions that you could take to reduce your emissions.

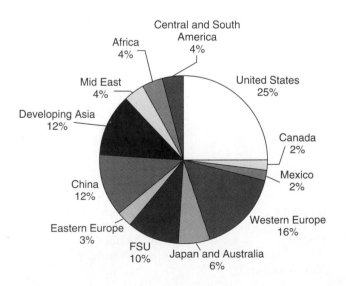

Figure 4.5 World CO_2 emissions, 1998. (FSU stands for Former Soviet Union.) (*Source*: Environmental Protection Agency.)

3. **Evolution of the atmosphere**
   Earth's early atmosphere looked very different from today's, and the story of its evolution provides an example of the close connections between the **atmosphere**, volcanism (**geosphere**), and life (**biosphere**).

   a. What is the age of the Earth?

   b. Briefly describe the **nebular hypothesis** for formation of Earth and the other planets (see p. 3-7 in this workbook and pp. 2–5 in *Earth Science*).

   c. Earth's first atmosphere probably consisted of the same elements as the solar **nebula** from which our planet was formed. Which elements would have been most abundant in this earliest atmosphere? (Hint: See pp. 2–3 in *Earth Science*.)

   d. The very first atmosphere was probably swept into space by the **solar wind**, a stream of particles emitted by the Sun. Earth's second atmosphere was formed when volcanoes transferred volatiles from the planet's interior to its surface in a process known as **outgassing** (p. 311 in *Earth Science*). This process continues today. Assuming that the gases emitted from modern volcanoes (see pp. 230–231 in *Earth Science*) are the same as those produced by outgassing early in our planet's history, what gases were present in Earth's second atmosphere?

   e. Were any elements that are present in the modern atmosphere (Table 4.3) missing from the second atmosphere?

   TABLE 4.3   Atmospheric compositions of Venus, Earth, and Mars.

	Venus (%)	Earth (%)	Mars (%)
Carbon dioxide (CO_2)	96	0.3	95
Nitrogen (N_2)	3.5	78	2.7
Oxygen (O_2)	Trace	21	0.13
Argon	0.007	0.9	1.6
Water vapor	0.1	Varies; 0-4	0.03

TABLE 4.4   Changing percentage of O_2 in Earth's atmosphere through time.

Millions of years before present									
4500	4000	3500	3000	2500	2000	1500	1000	500	0
CO_2 (%) 80	20	10	8	5	3	1	0.7	0.04	0.025
O_2 (%) 0	0	0	0	0	1	5	10	15	21

f. How does Earth's modern atmosphere compare to those of Venus and Mars?

g. Use the data in Table 4.4 to make a graph on Figure 4.6 of percent oxygen in the atmosphere as a function of time.

   1. About what percent of Earth history elapsed before oxygen began to accumulate in the atmosphere?

   2. When did the percentage of oxygen in the atmosphere reach about 50% of today's level?

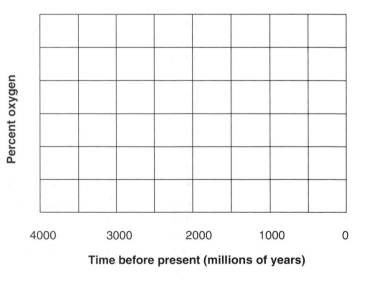

Figure 4.6 Graph of percent oxygen in Earth's atmosphere as a function of time.

Part IV. Concept Map About Global Warming

Fill in the question marks in the concept map shown in Figure 4.7. Add three new concepts and show how they link with other concepts. It would be helpful to complete the WebQuest on global warming before completing this concept map.

Part V. WebQuest - Global Warming: Scientific Fact or a Lot of Hot Air?

Introduction

Is Earth really getting warmer and, if so, are humans to blame? Even if the world is getting warmer, is it something we need to worry about? Might there actually be benefits to a warmer world? Global warming is perhaps the most "hotly" debated environmental issue of our time. The questions above are of much more than academic interest. There are economic and political prices to pay for curbing greenhouse gas emissions, and the many possible consequences of not doing so include rising sea level, severe weather, damage to ecosystems, and disease. How should we proceed? In this WebQuest, you will gather evidence to help you decide for yourself. You will use what you've learned to write a letter to the President of the United States in which you attempt to persuade him that your point of view is the correct one.

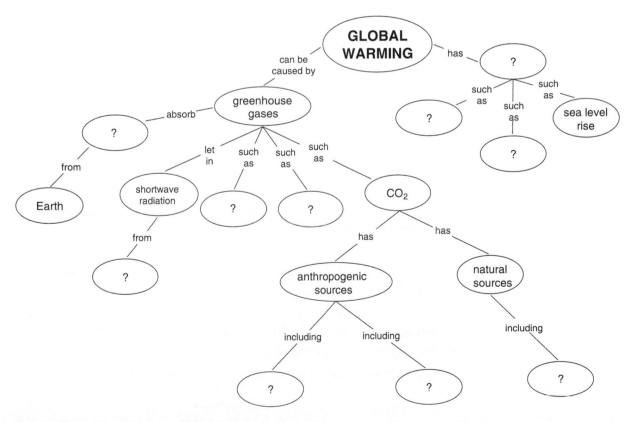

Figure 4.7 Concept map about global warming.

The Task

You've been hearing more and more about global warming, but nobody seems to agree, not even the scientists who study it. Is the world really in danger due to human-caused increases in atmospheric CO_2, or are the environmentalists a bunch of "Chicken Littles?" You've decided to find out what all of the controversy is about. Once you understand the two sides, you intend to write a letter to the President explaining why you think the United States should support or oppose international treaties to reduce CO_2 emissions.

The Process and Resources

Part I. Background

If your letter to the President is to be convincing, you had better know what you're talking about. The first step is to gather some information. Go to the Environmental Protection Agency's Global Warming Site at **http://www.epa.gov/globalwarming/** to find answers to the following questions.

1. What is the **greenhouse effect**?

2. List three **greenhouse gases**.

3. List three things that are known about global warming.

4. List three uncertainties about global warming.

Part II. Believers vs. Skeptics

In order to convince the President that your point of view is the correct one, you must understand how your opponents think.
To find out about the *believer's* point of view see:

- The Warming of the Earth (The Woods Hole Research Center)
  http://www.whrc.org/globalwarming/warmingearth.htm
- Union of Concerned Scientists
  http://www.ucsusa.org/warming/
- Environmental Defense Fund
  http://www.environmentaldefense.org/system/templates/page/issue.cfm?subnav=12

To find out about the *skeptic's* point of view see:

- Instant Expert Guide from the Heartland Institute
  http://www.heartland.org/studies/ieguide.htm
- Greening Earth Society
  http://greeningearthsociety.org/
- George C. Marshall Institute
  http://www.marshall.org/globalfax.html#fax1

Now, list three reasons for each position:

"Global warming is real":

"Global warming is hype":

Part III. The Letter

Write a one-page letter to the President of the United States in which you attempt to persuade him either that the United States must support international treaties to curtail CO_2 emissions, or that the United States cannot afford to do so. In your letter, you should demonstrate that you understand the opposing viewpoint and explain why your opinion is correct. Support your stance with facts, not opinion.

Learning Advice

- Did you use information from multiple sources?
- Did you compare information from different sources?
- Did you support your stance on global warming with facts?

Conclusion

What did you decide? Is global warming a real threat, or are there more important things to worry about?

Part VI. Summing Up the Sphere

Using information from the "Earth System Snapshots," "Earth as a System" boxes in your text, and your own ideas, describe interactions of the atmosphere with Earth's other components. Give at least two examples of each of the following interactions:

Atmosphere-Geosphere:

Atmosphere-Biosphere:

Atmosphere-Hydrosphere:

You may also think of examples that involve more than two spheres. Your examples:

Part VII. KWL Revisited

Go back to the KWL chart you filled in at the beginning of this module. Did you have any initial misconceptions about the topics you've explored? Go back to correct them. Do you have any unanswered questions?

Module 5
Cosmosphere

The Cosmosphere

In the preceding modules, you have explored Earth as a single **system** of interacting parts. But our home planet is not isolated from its surroundings; the **Earth system** is a system within other systems. Earth is part of the solar system, which is part of the Milky Way Galaxy, which is just one of billions of galaxies in the universe.

This module will focus on Earth's cosmic neighborhood, investigating what the study of other planets in the solar system can tell us about our own planet's past and future. By comparing Earth with other worlds, we gain a greater appreciation of what makes Earth uniquely suited for us and other life forms.

Learning Objectives

After completing the assignments in this module, you should be able to answer the following questions:

1. What can the study of **comets** tell us about the solar system?
2. How does the Sun interact with the Earth system?
3. How does the terrestrial atmosphere compare to those of Venus and Mars? What accounts for the very different atmospheres of these three planetary neighbors?
4. Why are Mars and Europa considered promising places to search for evidence of life beyond Earth?
5. How does Earth compare in size to the other planets of the solar system?
6. How big is the solar system?
7. How long would it take to travel to other planets and to the nearest star moving at: a. the highway speed limit? b. the speed of the fastest jet? c. the speed of light?
8. How are **terrestrial planets** different from **Jovian planets**? What accounts for the differences between these two groups of planets?
9. What terrestrial and extraterrestrial causes have been proposed for the extinction of the dinosaurs?

Part I. Getting Started: KWL Chart

Before you begin your study of Earth's place in the solar system, take a few moments to fill out Table 5.1. You will come back to it at the end of this module to see what you've accomplished, check if your questions have been answered, and ask any new questions arising from your investigations.

TABLE 5.1 KWL chart about the solar system.

K What I Know	W What I Want to Know	L What I Learned

Part II. Earth System Snapshots

Comets, Cosmic Clues, and Catastrophe

Themes: Interactions of Spheres; Scale

Although **comets** are small compared to other members of the solar system, they are big in excitement. People have always watched comets with wonder, and in ancient times they were often considered omens of evil fortune such as famine and war. In modern times we are less likely to see comets as portents of doom, but are no less awe-struck by the sight of a blazing comet. And comets, although small, may hold the answers to some really big questions about the origin and evolution of the solar system.

What are comets? In 1950, American astronomer Fred Whipple referred to them as "dirty snowballs" and the nickname has stuck. Comets are mixtures of ice (water, ammonia, methane, carbon dioxide, carbon monoxide), dust, and rock that are believed to come from the outer reaches of the solar system. Although their origin is not well understood, many comets are thought to come from the *Oort cloud*, a diffuse halo of debris surrounding our solar system at distances of 10,000–100,000 **Astronomical Units** (AU) from the Sun. We're talking way out there.[1] It's estimated that

---

[1] Recall that an Astronomical Unit or AU is defined as the average distance from Earth to the Sun: about 150 million km (93 million miles). Pluto is 39 AU from the Sun. The Oort cloud is more than 250 times farther than that!

the Oort cloud harbors billions or even trillions of comets, which are left-overs from the formation of the solar system. Other comets may come from the *Kuiper Belt*, which is closer to the Sun (a measly 30 to 100 AU away).

Unlike the planets, comets have highly elongated orbits and spend most of their time in the deep freeze of space beyond Pluto. Because comets from the Oort cloud may come from any direction and appear at any time (the very long period of very distant comets means that they may reappear only once in hundreds or thousands of years), astronomer David Levy has quipped that "comets are like cats. They have tails, and they do precisely what they want." When comets occasionally come into the inner solar system, the heat of the Sun causes the frozen gases to vaporize; this produces the **coma**, a bright, fuzzy cloud that surrounds the comet's solid core (its *nucleus*). Solar radiation and a steady flow of charged particles known as the **solar wind** push material away from the coma, producing a long *tail* that always points away from the Sun.

Comets are of interest to scientists for several reasons:

1. Because comets don't experience the heat and erosion that change planets, they may be the oldest, most unchanged bodies in the solar system. Comets are thus "time capsules" preserving material from the **nebula** that formed the Sun and planets;

2. Comets deliver water and other volatiles to planets and may have played a role in the formation of Earth's oceans and atmosphere;

3. Comets are rich in organic molecules and may have been involved in the origin of life on Earth;

4. Speeding comets and **asteroids** impact Earth and other planets with gigantic force, producing craters, causing major changes in climate, and severely disrupting the **biosphere**. Such an impact may have been responsible for the demise of the dinosaurs (Fig. 5.1).

In September 2001, NASA's Deep Space 1 probe passed just 2,200 km from the core of Comet Borrelly, capturing the best pictures to date of any comet's nucleus. In 2005, NASA plans to get a much closer look by crashing a 500 kg copper projectile into Comet Tempel 1 so that scientists can study pristine material excavated from its interior.

The Sun-Earth Connection

Themes: Interactions of Spheres; Scale; Cycles; Energy

Try to imagine how the Earth system would be different if there were no Sun. Without the Sun and its energy there would be no **photosynthesis**, no wind and rain, no seasons, and no humans or other life forms. In short, Earth would be a cold, dark, and lifeless place. This seems obvious but there are other, less apparent ways that the solar system's star player influences the Earth system.

The Sun may seem like a big featureless ball of light, but it is actually a dynamic body that, like the Earth, experiences cyclic changes. The best known of the cycles involves dark blotches on the Sun's surface called **sunspots**. Sunspots were noted by Chinese observers as long ago as 2,000 BC. They are relatively cool areas the size of Earth where the magnetic field is much stronger than elsewhere on the Sun. Sunspots are cooler than the surrounding **photosphere** because the concentrated magnetism inhibits

Figure 5.1 Painting by John Foster showing plesiosaurs and impacting asteroid. (*Source*: Photo Researchers, Inc.)

the upward flow of hot gas from the Sun's interior. Many centuries of observation have shown that the number of sunspots waxes and wanes in a cyclic fashion. About every eleven years the Sun reaches a peak period of activity known as "solar maximum," followed by a quiet interval called "solar minimum."

Sunspot groups are often associated with **solar flares** and **coronal mass ejections (CMEs)**. Solar flares are brief explosions that eject high-energy particles and radiation, while CMEs are much larger storms that violently thrust billions of tons of particles into space at speeds ranging from 20 to 2,000 kilometers per second. Solar flares and CMEs are forms of "space weather" that interact with Earth's atmosphere to produce dazzling **auroras** (Northern and Southern Lights). Space weather may disrupt communications and power systems on Earth, damage satellites, and endanger astronauts walking in space.

There have been many attempts to correlate solar cycles with global climate cycles on Earth, but this remains a hotly debated issue partly because Earth's climate is affected by many complex variables. During the

"Little Ice Age" of the mid-1600s to early 1700s, Western Europe experienced severe cold and a number of long winters. This was a period of very low sunspot activity (called the **Maunder Minimum**), but it is not known whether there was a causal link between the scarcity of sunspots and the cold, or whether it was just coincidence. Satellite measurements during the 1980s showed that incoming solar radiation decreased by only about 0.1 % from solar maximum to solar minimum, a variation that seems too small to drive temperature change on Earth. More importantly, no realistic causal mechanism has been proposed to link solar cycles with global climate cycles.

Goldilocks and the Three Planets

Themes: Interactions of Spheres; Cycles; Humans and the Earth System

If you were an intergalactic traveler visiting our solar system, your reaction to the temperatures of Earth and its two nearest neighbors might resemble that of Goldilocks sampling the three bears' bowls of porridge: Venus is too hot, Mars is too cold, but Earth is just right. You might think that this is because Venus is closest to the Sun and Mars is farthest away. But distance from the Sun isn't sufficient to explain why Earth alone is just right for the existence of liquid water, and therefore of life. In order to understand why these three worlds are so different, we need to explore their atmospheres.

Venus is so similar to Earth in its size, density, and chemical composition that it is often called our "sister planet." But you wouldn't want to visit Venus: in terms of temperature, it is more like Hell than like Earth. Its surface temperature reaches 475 °C (900 °F) — hot enough to melt lead! The Venusian atmosphere is dramatically different from ours: it's 97% carbon dioxide (a **greenhouse gas**) and so dense that atmospheric pressure is ninety times that on Earth's surface.

In contrast, the atmosphere of Mars is only 1% as dense as Earth's. Like Venus, Mars has an atmosphere that is mostly carbon dioxide, but the Martian atmosphere is so thin that there is little greenhouse warming. At the equator, Martian surface temperature averages about −60 °C (−76 °F) and it can plunge to −123 °C near the poles.

Venus, Earth, and Mars are planetary neighbors formed at the same time from the same solar **nebula**. Earth's very first atmosphere formed from the nebula and was rich in the light elements hydrogen and helium; the same was probably true of the primordial atmospheres of Venus and Mars. After their earliest atmospheres were swept away by the **solar wind**, secondary atmospheres formed on all three planets as a result of volcanic **outgassing** (see p. 311 in *Earth Science*). Why, then, are their present-day atmospheres so different? To understand how the Venusian, terrestrial, and Martian atmospheres evolved to their current states, we need to revisit the **carbon cycle** [see "Plate Tectonics and the Carbon Cycle" (Module 2), "Oceans and the Carbon Cycle" (Module 3), and "Does Life Control Climate?" (Module 4)].

Earth and Venus have about the same total amount of carbon dioxide, but whereas Venus's carbon dioxide resides in a thick, dense atmosphere, the Earth system's carbon dioxide is stored not only in the **atmosphere**, but also in the **geosphere, hydrosphere**, and **biosphere**. On Earth, carbon dioxide is added to the atmosphere by volcanism and metamorphism and

Figure 5.2 This image from the Mars Orbiter Camera shows a portion of the meandering canyons of the Nanedi Vallis, an ancient valley system of the southern highlands of Mars. A 200 m wide channel visible in the upper left suggests the valley was carved by water that flowed for an extended period of time. The canyon is about 2.5 km wide. (*Source*: NASA/JPL/Malin Space Science Systems.)

removed by chemical weathering and by organisms through photosynthesis and shell building; it also dissolves in the oceans. Venus has no oceans, probably because it's too close to the Sun for water vapor to condense. Without oceans or life to remove it, carbon dioxide has built up in the Venusian atmosphere. The situation on Venus provides a lesson for planet Earth when considering how humans may be causing global warming by adding carbon dioxide to our atmosphere.

Early in its history, Mars may have had a much thicker atmosphere than it does now and, as a result, the Red Planet may have once been warmer, perhaps warm enough for liquid water to flow on its surface (Fig. 5.2; see the following Earth System Snapshot, "Oases in the Solar System? The Search for Life Beyond Earth"). Mars may have lost most of its early, volcano-produced atmosphere in several ways. Because Mars is much smaller than Venus and Earth, it has less gravity to keep an atmosphere from being lost to space. As on Earth, carbon dioxide would have been removed from the atmosphere when it combined with water to make carbonate rocks. The small size of Mars led to quick cooling. With no active volcanism, there is no mechanism to cycle carbon dioxide from carbonate rocks back to the Martian atmosphere and Mars has a very small **greenhouse effect**. So, although you may be used to hearing about the negative consequences of human activities that enhance Earth's natural greenhouse effect by adding carbon dioxide to the atmosphere, most life as we know it would find Earth a very cold place if there were no greenhouse warming.

Oases in the Solar System? The Search for Life Beyond Earth

Themes: Interactions of Spheres; Energy

If you wanted to find life on other planets, where would you go? Because life as we know it requires liquid water, you'd do well to follow the water.

Of all of the planets and moons in the solar system, only Earth has liquid water at its surface. Current temperature and pressure conditions on Mars would cause liquid water on its surface to quickly freeze or evaporate, but there is evidence that, billions of years ago, water flowed on the

Red Planet. The Viking probes that landed on the Martian surface in 1976 did not find evidence of present-day life, but some scientists believe that what we should really be looking for is Martian *fossils*. Possible clues that Mars was once warmer and wetter and thus more hospitable for life include valleys resembling terrestrial stream systems (Fig. 5.2), channels that may have been formed by catastrophic floods, and layered sedimentary deposits.

In 1996, NASA scientists made an electrifying announcement: They had discovered fossils of a primitive life form deep inside "ALH84001," a meteorite discovered in Antarctica and believed to have come from Mars. While not everyone agrees that the microscopic worm-like features in the Martian meteorite actually represent fossilized bacteria, the discovery has renewed interest in the search for life beyond the home planet. Among the goals for current and future missions to Mars is the continued quest for Martian environments that may have once supported life.

Another body that is gaining attention in the search for extraterrestrial life is Europa, an icy moon of Jupiter that scientists believe may hide a liquid ocean beneath its icy crust. Detailed images taken by NASA's Galileo spacecraft show blocks of ice that appear to have broken away from a larger mass, possibly on a layer of liquid water or slush (Fig. 5.3). Although temperatures on the moon's surface average a chilly $-200\,°C$, it's possible that **tidal heating** due to the gravitational pull of Jupiter and neighboring moons could keep parts of Europa's ocean in the liquid state.

Of course, the current or past presence of liquid water doesn't guarantee the development of life, but the possibilities are intriguing and are helping to shape *astrobiology*, a relatively new field of science devoted to the study of the origin, distribution, and future of life in the universe. And astrobiologists aren't confining their search for extraterrestrial life to the "nearby" bodies of our own solar system. Since 1995, astronomers have used indirect evidence to discover dozens of planets in other solar systems. Using current technology, scientists have only been able to locate Jupiter-like gas giants, but the hunt is on for more Earth-like planets that may be more likely to harbor life. Is the Earth system unique among all planets in the Universe, or will further exploration uncover other worlds similar to our own?

Figure 5.3 Ice "rafting" on Europa. The size of the area shown is about 34 kilometers by 42 kilometers. Seen here are crustal ice plates up to 13 km across which have broken apart and moved, forming a pattern similar to that formed when pack ice on Earth's polar seas breaks apart during spring thaw. (*Source*: NASA Headquarters.)

Part III. Questions

A. Short Answer

1. **Scale in the universe: working with very large and very small numbers**

   To fully appreciate Earth's place in the universe, you need a sense of scale. When considering the size of planets and stars and the distances between them, we encounter numbers that are so immense that they are, well, "astronomical." For example, the mass of the Sun is 1,989,000,000,000,000,000,000,000,000,000 kilograms. Proxima Centauri, the closest star outside of our solar system, is 39,900,000,000,000 km away. That's almost 40 *trillion* km. Unless you are fond of writing zeros, there's got to be a better way. That's where *scientific notation* comes in. Using this shorthand notation, we can express the mass of the Sun as 1.989×10^{30} kilograms and the distance to Proxima Centauri as 3.99×10^{13} km. If it's been awhile since you thought about scientific notation, you might want to work through the following practice problems before tackling the Longer Answer questions below.

   a. *Using scientific notation to express really big numbers*

      Scientific notation is the expression of numbers as products of numbers between 1 and 10 times a power of ten. To write a number in scientific notation, place a decimal after the first digit. This gives you the *base number*. For example, in 525,000,000 the base number is 5.25. To find the *exponent*, count the number of places from the decimal to the end of the number. For 525,000,000, the exponent is 8. So 525,000,000 becomes 5.25×10^8.

      • For practice, convert the numbers in the Table 5.2 to scientific notation.

   b. *What about really small numbers?*

      The universe contains not only staggeringly large but also outrageously small things, such as atoms. The radii of atoms are measured in *picometers* (pm) (1 pm is equal to 0.000000000001 meters). To get rid of the zeros and express this number in scientific notation, move the decimal to the right until you have a number

TABLE 5.2 Expressing large numbers with scientific notation.

Number	Number in scientific notation
1,000,000,000,000,000,000 [length (in km) of our home galaxy, the Milky Way]	
93,000,000 [average distance (in mi) between Earth and the Sun]	
6,000,000,000,000 [a light-year: the distance (in mi) that light travels in a year]	
1,391,980 [diameter (in km) of the Sun]	

between 1 and 10. Count how many places the decimal was moved; this becomes the exponent in your scientific notation. In this example, moving the decimal 12 places to the right gives you the number 1.0, so 1 pm $= 1 \times 10^{-12}$ m.

- The radius of a hydrogen atom is about 37 picometers (pm) or 0.000000000037 m. Convert this number to correct scientific notation:

c. *Multiplication and division using scientific notation*

To multiply two numbers written in scientific notation, add their exponents.

Example: What is the product of 3×10^{12} and 2×10^{10}? (*Answer*: 6×10^{22}; $3 \times 2 = 6$ and $10^{12} \times 10^{10} = 10^{22}$.)

To divide one exponent by another, subtract the exponent on the bottom from the one on top. A simple example:

$$\frac{9 \times 10^5}{3 \times 10^3} = 3 \times 10^2 = 300$$

- For practice, multiply or divide as indicated and express your answer in scientific notation:

$$(3 \times 10^5) \times (2 \times 10^4) =$$

$$\frac{(4 \times 10^2) \times (2 \times 10^4)}{2 \times 10^3} =$$

$$\frac{10 \times 10^{12}}{2 \times 10^{25}} =$$

$$\frac{25 \times 10^{23}}{5 \times 10^8} =$$

2. Venus: Earth's Sister Planet?

Venus is often called Earth's sister planet. But, as you know, sisters often have their differences. Briefly discuss five ways that Venus is different from Earth.

3. a. Why does the Moon lack an **atmosphere**? (See p. 578 in *Earth Science*.)

b. How does interaction with the Moon affect the Earth's **hydrosphere**?

4. How would the Earth system be different if our planet were not tilted on its axis?

5. Keeping in mind the definition of **soil** given on p. 76 of *Earth Science*, do you think that you would find soil on the Moon?
Why or why not?

6. Would you expect to see arc-shaped chains of volcanoes on Mars? Why or why not? (Hint: See pp. 206–209 and pp. 249–252 in *Earth Science*.)

7. a. What type of volcano is found in Earth's Pacific Ring of Fire? (See p. 236 and pp. 249–252 in *Earth Science*.)

   b. Would you expect to find this type of volcano on Venus? Why or why not? (See pp. 583–584 in *Earth Science*.)

B. **Longer Answer**
   1. **A scale model of the solar system**
      a. *Size*
         To those of us who live on it, Earth seems like a pretty big and pretty important place. But Earth is not the center of the Universe; it's not even a particularly large planet when we look at the big picture. You probably already know that the solar system is really BIG, but just how big? The solar system is truly immense when compared to anything we encounter in our everyday lives, and our minds are easily boggled by the sizes of objects in the solar system and the distances between them. A good way to understand sizes and distances in space is to construct a *scale model* using familiar objects. In this exercise, you will imagine a model solar system in which the Sun is the size of a basketball. The model will help you to picture how Earth compares in size to the Sun and to other planets. It will also provide a concrete way for you to grasp the magnitude of distances within the solar system.
         1. *Testing prior knowledge*
            Before you begin, test your pre-existing notions about scale in the solar system. In a model in which the Sun is scaled to the size of a basketball, which of the following objects would most accurately portray the size of Earth?
            a. Golf ball
            b. Penny
            c. Peppercorn
            d. Pinhead
            e. Speck of dust
         2. *Constructing the model*
            The first step in constructing your model is to determine the *scaling factor* that will be used. To do this, you need to compare the diameter of the basketball to that of the real Sun. The diameter of a basketball is about 23.3 cm or 0.233 m. The diameter of the Sun is 1,391,980 km or 1,391,980,000 m. Thus, the ratio of the size of the model Sun to that of the real Sun is 0.233:1,391,980,000 or 1:6 billion (or very close). This means

that every unit in your model represents 6 billion actual units. For example, 1 cm on your model Sun represents 6 billion cm on the actual Sun. In other words, the model shrinks the solar system to one 6-billionth of its actual size. To scale other objects in the model:

Scaled diameter =

$$\frac{\text{Actual diameter (km)}}{6{,}000{,}000{,}000} = \frac{\text{Actual diameter (km)}}{6 \times 10^9}$$

For example, to figure out the scale diameter of Mercury:

$$\text{Scaled diameter} = \frac{4{,}880 \text{ km}}{6 \times 10^9} = \frac{4.88 \times 10^3}{6 \times 10^9}$$

$$= 0.81 \times 10^{-6} \text{ km} \times \frac{10^5 \text{ cm}}{1 \text{ km}} =$$

$$0.81 \times 10^{-1} \text{ cm} = 0.081 \text{ cm}.$$

Go ahead and calculate the scale diameters of the other planets and fill in the appropriate column in Table 5.3. *Show your work.

Venus:

Earth:

Mars:

Jupiter:

TABLE 5.3 A scale model of the solar system where the Sun is the size of a basketball.

Object in Solar System	Diameter (km)	Scale Diameter (cm)	Object in model	Mean Distance from the Sun (millions of km)	Mean Distance from the Sun (AU)	Scale Distance from Sun (m)
Sun	1,391,980	23.3	basketball	-		
Mercury	4,880	0.08		58	0.39	
Venus	12,100			108	0.72	
Earth	12,800			150	1.00	25
Mars	6,800			228	1.52	
Jupiter	143,000			778	5.20	
Saturn	121,000			1427	9.54	
Uranus	51,800			2870	19.18	
Neptune	50,530			4497	30.06	
Pluto	2,300			5900	39.44	

TABLE 5.4 Approximate sizes of some common objects.

Object	Approximate diameter
basketball	23.3 cm
softball	11.2 cm
grapefruit	10 cm
golf ball	4.5 cm
walnut or pingpong ball	3.8 cm
quarter	2.5 cm
penny	1.9 cm
peanut	7.5 mm
peppercorn	2 mm
pinhead	1 mm
poppy seed	0.3 mm
speck of dust	<0.3 mm

Saturn:

Uranus:

Neptune:

Pluto:

Now, select an object from Table 5.4 that reasonably represents each planet at this scale and fill in the appropriate column in Table 5.3.

b. *Distance*
If your model is to accurately portray the sizes of the planets and their distances from the Sun, how big does the model need to be? The real distance between the Sun and Earth is 150 million km. But trying to measure astronomical distances in kilometers is a bit like using millimeters to measure the height of Mt. Everest. One way to deal with very large distances in space is to devise new units of measurement such as the **Astronomical Unit** (AU for short), which is defined as the distance between the Sun and Earth. We can then compare other distances to the Sun-Earth distance. On the 1:6 × 10⁹ scale of your model, how far is one AU?

$$1 \text{ A.U.} = \frac{150 \times 10^6 \text{ km}}{6 \times 10^9} = 25 \times 10^{-3} \text{ km} \times \frac{1000 \text{ m}}{1 \text{ km}} = 25 \text{ m}.$$

Use the fact that 1 AU = 25 m in your scale model to determine the scale distances of the other planets from the Sun. Add your answers to Table 5.3.

To summarize, if you shrank the solar system to one 6-billionth of its actual size, the Sun would be the size of a basketball and Pluto would be the size of a poppy seed at a distance of 975 meters (about 0.6 miles) away!

2. **Cosmic Traveler**

   Another way to appreciate the vast size of the universe is to figure out how long it would take to travel to distant places in the solar system and beyond. If you are a science fiction fan, you may take interstellar travel for granted, but in reality there is no such thing as warp speed.

   a. To date, the Moon is the only extraterrestrial world to be visited by humans. It has an average distance of about 400,000 kilometers from Earth. How long would it take you to drive to the moon at a speed of 100 km/hr?

$$400{,}000 \text{ km} \times \frac{1 \text{ hour}}{100 \text{ km}} = 4{,}000 \text{ hours} \times \frac{1 \text{ day}}{24 \text{ hours}} = \ \sim 167 \text{ days}$$

   b. Using the values given in Table 5.5, calculate travel times to the Moon if you were able to travel at the following speeds. *Show your work.

   A jet:

   A rocket:

   Light (the fastest thing in the universe):

   c. What if you are more adventurous and would like to visit Jupiter? Calculate travel times to Jupiter at various speeds and fill in the appropriate column in Table 5.5. *Show your work.

   Time to Jupiter traveling at the speed of:
   A jet:

TABLE 5.5 Travel times to distant bodies in the solar system and beyond.

Approximate Speed		Time to the Moon (average 400,000 km away)	Time to Jupiter (average ~ 630 million km away)	Time to nearest star outside solar system (~ 40 trillion or 40 × 10^{12} km away)
Freeway speed	100 km/hour	~ 167 days		Too long!
Jetliner speed	1,000 km/hour			Too long!
Rocket speed	40,000 km/hour			Too long!
Speed of light	300,000 km/second			

A rocket:

Light:

d. You don't want to be confined to the solar system? How much time should you set aside for your trip to the nearest star outside the solar system (Table 5.5)? Let's assume that you want to skip the slow stuff and travel at the speed of light. How long will it take you? *Show your work.

Where's that warp drive when you need it?

3. **Planet Patterns**
   a. Using the planetary data provided in Table 5.6 (note that some of the values in this table differ slightly from those in Table 21.1 of *Earth Science*), make a graph on Figure 5.4 of mean density vs. orbital distance. Label each planet on your graph. Describe any relationship you see between a planet's mean density and its distance from the Sun.

   b. Use the information given in Table 5.6 to label each planet plotted on Figure 5.5. Do you see any relationship between a planet's size and its distance from the Sun?
   Explain.

   c. The planets can in general be grouped into two sets based on density and size (Table 5.7).

TABLE 5.6 Properties of the planets in the solar system. Eq. Stands for equatorial.

Property	Mercury	Venus	Earth	Mars	Jupiter	Saturn	Uranus	Neptune	Pluto
Mass (10^{24} kg)	0.3302	4.869	5.975	0.6419	1,898.65	568.46	86.83	102.43	0.0125
Eq. radius (km)	2,439	6,052	6,378	3,393	71,492	60,268	25,559	24,766	1,137
Mean density (kg/m^3)	5,427	5,204	5,520	3,933	1,326	687	1,318	1,638	2,050
Orbital distance (10^6 km)	57.9	108.2	149.6	227.9	778.3	1,427.0	2,869.6	4496.6	4,913.5
Orbital period (days)	87.969	224.7	365.25	686.98	4330.5	10,747	30,588	59,800	90,591

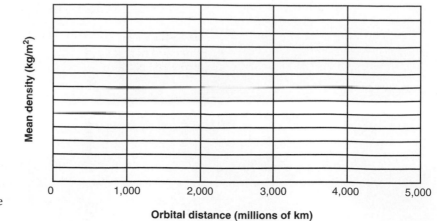

Figure 5.4 Graph of mean density vs. orbital distance for planets of the solar system.

1. In Table 5.7, list the planets that fall into each category.

2. To which category did you assign Pluto?
   Explain your reasoning.

d. Referring to Tables 5.6 and 5.8,
   1. Which two materials do you think are most abundant in the terrestrial planets?
      Explain your reasoning.

   2. Which planet would float in water?

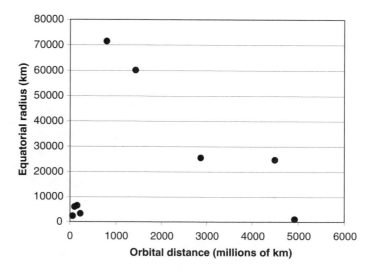

Figure 5.5 Graph of equatorial radius vs. orbital distance for planets of the solar system.

TABLE 5.7 Terrestrial vs. Jovian planets.

Small and high density (terrestrial planets)	Large and low density (Jovian planets)

TABLE 5.8 Densities of
planetary materials.

Material	Density
Air	1.2 kg/m^3
Water or Ice	1,000 kg/m^3
Typical Rocks	3,000 kg/m^3
Metal at High Pressure	10,000 kg/m^3

e. Orbital period
   1. Referring to Figure 5.6, how would you describe the relationship between a planet's orbital period and its distance from the Sun?

   2. The orbital period of the Earth is about 365 days. What is this called?
   3. Refer to Table 5.6 to answer the next three questions.
      a. How long is a year on Mercury?
      b. How old are you in Saturn years? *Show your work.

      c. How many Earth years go by before one Pluto year has passed? *Show your work.

f. Atmospheres (See Table 5.9)
   1. Which two planets have similar atmospheric compositions?
   2. What makes these two atmospheres very different?
   3. Which gas is present in Earth's atmosphere, but nearly absent in the atmospheres of Venus and Mars?

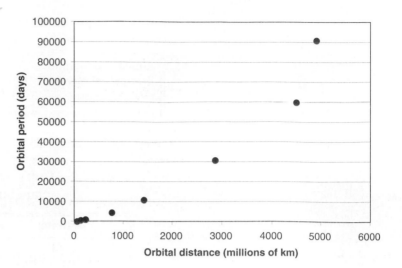

Figure 5.6 Graph of orbital period vs. orbital distance for planets of the solar system.

What do you think explains its presence on Earth? (Hint: What does Earth have that Venus and Mars both lack?)

4. Based on the data in Table 5.9, how would you describe the effect of increasing atmospheric pressure on the variability of a planet's surface temperature?

5. Note that even though Mercury is closer to the Sun, Venus is somewhat hotter. Why?

Part IV. Concept Map About the Solar System

Construct a concept map about the solar system. Include the terms listed below on your map. Add at least five terms of your choosing and show appropriate links to the other concepts.

TABLE 5.9 Comparison of the atmospheres and surface temperatures of Mercury, Venus, Earth, Mars, and the Moon.

Planet or Body	Gases (% by volume)			Mean Distance from Sun (millions of km)	Surface Temperature (range)	Surface Atmospheric Pressure (bars)
	N_2	O_2	CO_2			
Mercury	0	trace	0	58	−173° to 427°C	10^{-15}
Venus	3.5	< 0.01	96.5	108	475°C (small range)	92
Earth	78.01	20.95	0.03	150	−40° to 75°C	1.014
Earth's Moon	0	0	0	Similar to Earth	−173° to 130°C	Essentially 0
Mars	2.7	1.3	95.32	228	−120° to 25°C	0.008

Planets	Comets	Asteroids
Terrestrial	Jovian	Size
Density	Rock	Metal
Gas	Ice	Mars
Atmosphere	Helium & hydrogen	Carbon dioxide
Oxygen	Earth	Venus
Jupiter	Neptune	Rings
Phobos	Io	Titan
Ceres	Hale-Bopp	

Part V. WebQuest - What Killed the Dinosaurs?

Introduction

In this workbook, you have studied the interacting parts of the **Earth system** and have seen that the **biosphere** impacts and is impacted by the **geosphere, atmosphere**, and **hydrosphere**. At several times through Earth history, there have been relatively sudden decreases in the diversity of life. Such widespread perturbations of the biosphere are known as **mass extinctions**. The most famous mass extinction, which involved the demise of the dinosaurs, occurred about 65 million years ago at the boundary between the *Cretaceous* and *Tertiary* periods (the *KT boundary*; see Box 11.3 on p. 325 of *Earth Science*). This extinction involved not only the dinosaurs, but also many species of marine organisms. What changes in the Earth system could have disturbed the **biosphere** so much that more than 50% of living organisms died out? Was it a **comet** or an **asteroid** crashing into our planet, or did it involve changes in one or more of the terrestrial spheres?

The Task

In this WebQuest, you will research possible causes of dinosaur extinction and then write a position paper explaining which hypothesis or hypotheses you favor and why.

The Process and Resources
Part I. Gathering information

Use the following Web sites to identify *four hypotheses* for what killed the dinosaurs. List the hypotheses in Table 5.10 and briefly summarize the evidence that supports each.
- **http://www.bbc.co.uk/beasts/whatkilled/theories/**
- **http://www.bbc.co.uk/education/darwin/exfiles/cretaceous.htm**
- **http://www.ucmp.berkeley.edu/diapsids/extinction.html**
- http://www.enchantedlearning.com/subjects/dinosaurs/ extinction/Other.html
- **http://filebox.vt.edu/artsci/geology/mclean/Dinosaur_Volcano_Extinction/**

Part II. Position Paper

Write a brief (about one page) essay explaining which hypothesis or hypotheses you favor and why. Support your position with facts from the Internet sites that you have visited.

TABLE **5.10** Four hypotheses for what killed off the dinosaurs and many other species 65 million years ago.

Hypothesis	Evidence
1.	
2.	
3.	
4.	

Learning Advice

- Did you use information from multiple sources?
- Did you compare information from different sources?
- Did you support your favored hypothesis about what killed the dinosaurs with facts?

Conclusion

What are some processes and events that can perturb the Earth system enough to cause widespread extinction of organisms? Which of these processes or events caused the mass extinction at the KT boundary?

Part VI. Summing Up the Sphere

Using information from the "Earth System Snapshots," "Earth as a System" boxes in your text, and your own ideas, describe interactions of the **cosmosphere** with Earth's other components. Give at least two examples of each of the following interactions:

Cosmosphere-Geosphere:

Cosmosphere-Atmosphere:

Cosmosphere-Biosphere:

Cosmosphere-Hydrosphere:

You may also think of examples that involve more than two spheres. Your examples:

Part VII. KWL Revisited

Go back to the KWL chart you filled in at the beginning of this module. Did you have any initial misconceptions about the topics you've explored? Go back to correct them. Do you have any unanswered questions?

Finally, go back to the *very first KWL chart you completed* (p. 1-4). Did you have any initial misconceptions about planet Earth? Go back to correct them. Do you have any unanswered questions?

Glossary

Anthropogenic   Caused or influenced by humans.

Biological pump   The transfer of CO_2 from shallow to deep ocean water as a result of photosynthesis near the surface and subsequent settling and deposition of organic material.

Carbon cycle   Describes how carbon and carbon compounds are transferred between the atmosphere, hydrosphere, biosphere, and geosphere. The **long-term carbon cycle** operates on time scales of more than 100,000 years and involves such processes as volcanism, metamorphism, and chemical weathering. The **short-term carbon cycle** operates on time scales of days to thousands of years, with the major processes being **photosynthesis** and **respiration**.

Coronal mass ejection   A giant bubble of gas ejected from the Sun's corona that can carry as much as 10 billion tons of solar gas and may travel at speeds of 2,000 kilometers per second.

Cosmosphere   The universe surrounding the **Earth system**.

Deforestation   The conversion of forested to nonforested land (such as pasture, croplands, urban areas). Deforestation enhances the natural greenhouse effect because burning or decomposition of wood releases carbon dioxide and the felled trees no longer remove carbon dioxide by photosynthesis.

Earth system   A unified set of interrelated components (atmosphere, hydrosphere, geosphere, biosphere, and cosmosphere) which interact and function as an integrated whole.

Feedback   Factors that amplify (**positive feedback**) or counteract (**negative feedback**) an initial change in a system.

Gaia hypothesis   States that Earth is a self-regulating system in which living things play an integral role.

Geosphere   The solid part of the **Earth system** extending from the core to the crust and including minerals, rocks, and soil.

Greenhouse gas   A gaseous component of the atmosphere that warms a planet's surface by absorbing infrared radiation and re-radiating some of it back towards the surface. Greenhouse gases include carbon dioxide, methane, nitrous oxide, chlorofluorocarbons, and water vapor.

Long-term carbon cycle   See **Carbon cycle**.

Mass extinction   A catastrophic and widespread (often global) event where major groups of species become extinct in a relatively short interval of geologic time compared to the normal, continuous low-level extinction that has occurred throughout most of the history of life.

Maunder Minimum   A period between 1645 and 1715 (70 years), when (for unknown reasons) sunspots were unusually scarce. Corresponds to the "Little Ice Age," when temperatures in Europe were unusually cold.

Negative feedback   See **Feedback**.

Nitrogen cycle   Describes the movement of nitrogen in its many forms between the atmosphere, hydrosphere, biosphere, and geosphere.

Paleoclimate   Climate of the distant past, particularly before historical records.

Paleotemperature   Temperature of the distant past, particularly before historical records.

Parts per million (ppm)   A unit used to express very dilute concentrations of a substance. Just as 1% is equal to one part in 100 total parts, 1 ppm is equal to one part in 1,000,000 total parts. For example, the concentration of CO_2 in the atmosphere is 360 ppm, meaning that only 360 of every 1 million parts of Earth's atmosphere are carbon dioxide.

Phytoplankton   Tiny plants, especially algae, that drift in the surface layers of a body of water where there is sufficient penetration of sunlight to allow **photosynthesis**.

Positive feedback   See **Feedback**.

Proxy data   A proxy is a substitute for the real thing. In studies of **paleoclimate**, proxy data from tree rings, ice cores, fossil pollen, corals, and other natural recorders of climate variability fill in for direct measurements and extend our understanding of climate far beyond the instrumental record of 140 years.

Reservoir   In cycles such as the hydrologic and carbon cycles, a natural or artificial storage place for a substance.

Residence time   A measure of the average time that an atom or molecule of a given substance spends in a **reservoir**.

Respiration   The chemical process (the reverse of **photosynthesis**) by which plants and animals release the energy stored in food. Carbon dioxide is a product of this process.

Short-term carbon cycle   See **Carbon cycle**.

Specific heat   The amount of energy needed to raise the temperature of 1 gram of a substance by $1°$ Celsius.

Steady-state equilibrium   A type of balance in which the average condition of a **system** remains unchanged over time.

Tidal heating   Frictional heating of a planetary body's interior due to flexure caused by the gravitational pull of a larger nearby object or objects.

Vog   Smog of volcanic origin. Formed when sulfur dioxide gas and other pollutants emitted from volcanoes react with oxygen and moisture in the atmosphere.

References

Module 1: Introduction to the Earth System

Beginning

Sussman, Art, 2000, *Dr. Art's Guide to Planet Earth*, Chelsea Green Publishing Co., White River Junction, VT, 122 p.

Web Resource

Hamilton, Calvin J., 2001, Earth Introduction, in *Views of the Solar System*, **http://www.solarviews.com/eng/earth.htm** [This Web site has great images of and movies about planet Earth.]

Advanced

Kump, Lee R., Kasting, James F., and Crane, Robert G., 1999, *The Earth System*, Prentice-Hall, Upper Saddle River, NJ, 351 p.

Mackenzie, Fred T., 1995, *Our Changing Planet* (2nd ed.), Prentice-Hall, Upper Saddle River, NJ, 486 p.

Ruddiman, William F., 2001, *Earth's Climate, Past and Future*, W.H. Freeman Co., New York, 465 p.

Web Resource

Martin, Laurie, 1997, An Introduction to Feedback, **ftp://sysdyn.mit.edu/ftp/sdep/Roadmaps/RM2/D-4691.pdf.** This site works only with Microsoft Internet Explorer.

Module 2: Geosphere

Geosphere 1: Earth Materials

Beginning

Web Resource

United States Geological Survey, 1997, Acid Rain and Our Nation's Capital: **http://pubs.usgs.gov/gip/acidrain/**

Geosphere 2: Earth Dynamics

Beginning

Abbott, P.L., 1999, *Natural Disasters* (2nd ed.), WCB/McGraw-Hill, New York, 397 p.

Kious, J.W. and Tilling, R. I., 1996, *This Dynamic Earth*, U.S. Government Printing Office, Washington, D.C. Available on-line at **http://pubs.usgs.gov/publications/text/dynamic.html**

United States Geological Survey, in preparation, *This Dynamic Planet: A Teaching Companion*.

Web Resources

Bigalow Laboratory for Ocean Sciences, 2001, Data from Ocean Buoys, **http://www.bigelow.org/virtual/index_buoy.html** [This site features information about ocean waves, including an animation of tsunami formation.]

Michigan Technical University, 2001, Lahars, **http://www.geo.mtu.edu/volcanoes/hazards/primer/lahar.html**

National Aeronautics and Space Administration (NASA), 2001, Classroom of the Future, The Carbon Cycle, **http://www.cotf.edu/ete/modules/carbon/efcarbon.html**

United States Geological Survey (USGS):

- 2001, Volcano Hazards Program, **http://vulcan.wr.usgs.gov/Vhp/framework.html**
- 2000, Selected Case Studies: Hazardous Volcanic Activity, **http://volcanoes.usgs.gov/Hazards/ What/hazards.html#Studies**
- 2001, Tsunamis & Earthquakes, **http://walrus.wr.usgs.gov/tsunami/index.html**

Advanced

Berner, Robert A., 1999, A New Look at the Long-term Carbon Cycle: *GSA Today,* v. 9, pp. 1–6.

Ruddiman, William F., 2001, *Earth's Climate, Past and Future,* W.H. Freeman Co., New York, 465 p.

Module 3: Hydrosphere

Beginning

Web Resources

National Aeronautics and Space Administration (NASA), 2001, Classroom of the Future, The Carbon Cycle, **http://www.cotf.edu/ete/modules/carbon/efcarbon.html**

National Aeronautics and Space Administration (NASA), 1999, Clouds and the Energy Cycle, Fact Sheet NF-207, **http://eospso.gsfc.nasa.gov/ftp_docs/Clouds.pdf**

United Nations Environment Progamme, 1999, Global Environment Outlook 2000, **http://www.grida.no/geo2000/**

United Nations Environment Progamme, 1993, Oceans and the Carbon Cycle, Information Unit on Climate Change Fact Sheet 21, **http://www.unep.ch/iucc/fs021.htm**

Advanced

Berner, Elizabeth K. and Berner, Robert A., 1996, *Global Environment*: *Water, Air, and Geochemical Cycles,* Prentice-Hall, Upper Saddle River, NJ, 376 p.

Freeze, R. Allan and Cherry, John A., 1979, *Groundwater,* Prentice-Hall, Upper Saddle River, NJ, 604 p.

Schlesinger, William H., 1997, *Biogeochemistry: An Analysis of Global Change* (2nd ed.), Academic Press, New York, 588 p.

Module 4: Atmosphere

Beginning

McGuire, Thomas, 1993, The Gaia Nineties: *The Science Teacher,* v. 60, no. 6, pp. 30–35.

Web Resources

Camp, Vic, 2001, Climate Effects of Volcanic Eruptions, **http://www.geology.sdsu.edu/ how_volcanoes_work/climate_effects.html**

National Aeronautics and Space Administration (NASA), 2001, Classroom of the Future, The Carbon Cycle, **http://www.cotf.edu/ete/modules/carbon/efcarbon.html**

National Air and Space Museum, 1999, Exploring the Planets: Planetary Atmospheres, **http://www.nasm.edu/ceps/etp/compare/atmos/atmos.html**

National Oceanographic and Atmospheric Administration (NOAA), 2001, Paleoclimatology Program, **http://www.ngdc.noaa.gov/paleo/paleo.html**

Stahle, David W., 2001, Dead Men's Tales, Climatology: Lessons From the Past and the Reality of Global Warming: Scientific American Frontiers Web Feature: **http://www.pbs.org/saf/1203/features/climatology.htm**

United States Geological Survey (USGS), 2000, Sea Level and Climate, Fact Sheet 002-00, **http://pubs.usgs.gov/factsheet/fs2-00/**

Advanced

Ruddiman, William F., 2001, *Earth's Climate*: *Past and Future*: W.H. Freeman, New York, 465 p.

Smil, Vaclav, 2000, *Cycles of Life*: *Civilization and the Biosphere*, Scientific American Library, W. H. Freeman and Company, 240 p.

Stahle, David W., 1998, The Lost Colony and Jamestown Droughts: *Science,* v. 280, pp. 564–567.

Vitousek, P. M., Aber, J., Howarth, R.W., Likens, G.E., Matson, P.A., Schindler, D.W., Schlesinger, W.H., and Tilman, G.D., 1997, Human Alteration of the Global Nitrogen Cycle: Causes and Consequences: *Issues in Ecology,* v.1, pp. 1–15. Available on-line at **http://esa.sdsc.edu/tilman.pdf**

Web Resources

National Aeronautics and Space Administration (NASA), 1998, Polar Ice, Fact Sheet NF-212, **http://eospso.gsfc.nasa.gov/ftp_docs/Polar_Ice.pdf**

Module 5: Cosmosphere

Beginning

Cherif, Abour H., and Adams, Gerald E., 1994, Planet Earth: Can Other Planets Tell Us Where We Are Going?: *The American Biology Teacher,* v. 56, no. 1, pp. 26–37.

Kargel, J. S., and Strom, R. G. 1996, Global Climatic Change on Mars: *Scientific American,* v. 275, no. 5, pp. 80–88.

McKay, Christopher P., 1997, Looking for Life on Mars: *Astronomy* (August), pp. 38–43.

Marcy, Geoffrey W. and Butler, R. Paul, 1998, Giant Planets Orbiting Faraway Stars: *Scientific American Quarterly,* v. 9, no. 1, pp. 10–15. Available on-line at **http://www.sciam.com/specialissues/ 0398cosmos/0398marcy.html**

Nadis, Steve, 2001, Searching for the Molecules of Life in Space: *Sky & Telescope,* v. 103, no. 1, pp. 32–47.

Phillips, Cynthia, 1999, Planets and Satellites Curriculum Module, Galileo Solid State Imaging Team, Education and Public Outreach (NASA).

Web Resources

National Aeronautics and Space Administration (NASA):

- Comets (Stardust Project Homepage), **http://stardust.jpl.nasa.gov/comets/**
- Asteroid and Comet Impact Hazards, **http://impact.arc.nasa.gov/**
- Exobiology: The Search for Life on Mars, **http://cmex-www.arc.nasa.gov/CMEX/data/SiteCat/sitecat2/exobiolo.htm**

Advanced

Kasting, J.F., Toon, O.B., and Pollack, J.B., 1988 (Feb.), How Climate Evolved on the Terrestrial Planets: *Scientific American,* v. 258, no. 2, pp. 90–97.